Maryon Stewart studied preventative dentistry and nutrition at the Royal Dental Hospital in London and worked as a counsellor with nutritional doctors in England for four years. She set up the PMT Advisory Service at the beginning of 1984 which has subsequently provided help to thousands of women all over the world. In 1987 she launched the Women's Nutritional Advisory Service, which now provides broader help to women. Maryon is the author of the best-selling books *Beat PMT Through Diet*, *The Vitality Diet* and the co-author of *The PMT Cookbook*. She has a weekly radio programme on health and nutrition, has co-written several medical papers and written articles for many glossy magazines including *Marie Claire*, *Cosmopolitan*, *Chat*, *Woman's Journal*, and *Health & Fitness*. She has also appeared on several popular TV magazine shows, like *TV AM*, *The Miriam Stoppard Health & Beauty Show*, *This Morning*, *Pebble Mill*, Channel Four *Streetwise* and Yorkshire and TVS's *Help Yourself* programmes. She is married to Dr Alan Stewart, and together they live in Lewes, in Sussex, with their four children.

By the same author

Beat PMT Through Diet
Beat PMT Cookbook
The Vitality Diet

BEAT SUGAR CRAVING

The Revolutionary 4-Week Diet

MARYON STEWART

VERMILION
LONDON

Roses are red
Violets are blue
Sounds all very fine
But how are you?

To sugar cravers the world over

Published in 1992 by Vermilion
an imprint of Ebury Press
Random Century House
20 Vauxhall Bridge Road
London SW1V 2SA

Text Copyright © Maryon Stewart 1992
Cartoons Copyright © Mike Gordon 1992

Catalogue record for this book is available
from the British Library.

ISBN 0 85223 915 7

Designed by: Bob Vickers
Edited by: Helen Southall

Typeset by Hope Services (Abingdon) Ltd.
Printed and bound by
Mackays of Chatham, PLC,
Kent.

Contents

Acknowledgements   6

Introduction   7

PART I – THE SUGAR CRAVING SYNDROME

 1 Sugar the World Over   11
 2 Sweet History   16
 3 What the Experts Say   20
 4 The Ad-Man's Delight   27
 5 The Sugar Craving Syndrome   32
 6 Confessions of 'Chocoholics'   38
 7 Sweet Nothing – The Nutritional Value of Sugar   44
 8 All About Low Blood Sugar   49
 9 A Sweet Tooth   59
10 Artificial Sweeteners   68
11 Sugar – The Reward?   72
12 Mental Health in the Balance – Eating Disorders   82

PART II – BEATING THE CRAVINGS

13 The Good News   85
14 Making a Workable Plan   90
15 The 30-Day Diet   105

Recommended Reading   182
Specialised Food and Nutritional Supplement Suppliers   184
Useful Addresses   186
Index   189

Acknowledgements

I would like to acknowledge all the members of both the medical and dental professions who have been involved in researching the subject of sugar and its effect on health. The scientific groundwork they have provided has been of immense value. My special thanks also go to the many patients who have willingly offered to share their case histories, no matter the personal cost, in order to help others.

Untold gratitude is also due to two wonderful helpers without whose contributions my task would have been a great deal more onerous. Dr Alan Stewart contributed to Part I of the book and has been a constant technical support throughout, and Sarah Tooley, Senior Nurse at the Women's Nutritional Advisory Service (WNAS), contributed greatly to the menu and recipe section in Part II.

Thanks are also due to other dedicated members of the WNAS team whose help with research and administration has been invaluable: Jenny Tooley, Michèle Apsey, Jayne Tooke, Jenny Parkhouse and Karen Stone. Additionally, thanks should go to three honorary staff members, Brian Hendry, and Phoebe and Chesney Stewart, for their TV research.

I am, as ever, very grateful to Lavinia Trevor for her constant guidance, and to Rowena Webb and the team at Ebury Press for their patience and professionalism.

I cannot end without thanking my nanny, Jane Worfolk, for happily looking after my children while I was putting in the extra hours on this book, nor without thanking the children themselves, Phoebe, Chesney, Hester and Simeon (who was actually conceived at the same time as the book), for their patience, understanding and constant sense of humour.

Maryon Stewart

Introduction

As RECESSION BITES in the early 1990s, you may well be spending less on holidays, cars, clothes and evenings out, but the chances are that you are spending more than ever on chocolates and confectionery. While sales of luxury goods continue to slump, sales of chocolate and confectionery are at an all-time high, suggesting that either the 'chocoholics' amongst us have increased their consumption, or that more and more of us are fast becoming comfort eaters. If you have ever had the desire to eat half a packet of chocolate biscuits in one sitting, or half a dozen chocolate bars in quick succession, you will know how difficult it is under normal circumstances to say no to a craving for something sweet. Regardless of how rational you may usually be, the compulsion to indulge is sometimes overwhelming.

Cast aside any guilt you may be feeling at this point; you are certainly not alone. During the eight years that the Women's Nutritional Advisory Service (WNAS) has been operating, I have come to realise just how common a problem the Sugar Craving Syndrome is. Cravings can vary from a desire to eat a chocolate bar daily, to the compulsion to scoff several packets of chocolate biscuits at one sitting. If your desire for sweet food frequently overtakes your will-power, the problem needs to be addressed, unless, of course, you consider you have a perfect figure, are in optimum health, and have a complexion like peaches and cream!

The Sugar Craving Syndrome has been pretty much ignored by the medical profession. In fact, it has never been regarded as a genuine syndrome, needing treatment or advice. Consequently, people have been left to battle alone with their 'addiction'. To illustrate how serious a problem it really is, here are a few experiences that some of our grateful patients have been willing to share.

Pauline Hutchins remembers how her sugar cravings affected her:

> 'The worst time was when I had no car available to me. I lived in the country, and there were no shops nearby. In desperation I made two dozen jam tarts in an attempt to stop my cravings. I ate them all while they were still hot. My family never saw them. But I still wanted chocolate afterwards.'

The 'chocoholic', or severe sugar craver, is in a similar situation to the alcoholic or heavy smoker. These are not just unpleasant habits, they are very real addictions. It is easy to tell a 'chocoholic' to cut down on his or her sweet food intake but, as you may be aware, it is not so easy to put it into practice.

Michael had been eating several chocolate bars a day for as long as he could remember:

> 'To be perfectly honest, I'd rather go without food and have chocolate instead. I'd stock up with at least four bars of chocolate on my way to work. During the course of the day I'd steadily work my way through them. Before I met my fiancée, on lonely evenings, I would sit and eat four or five bars in quick succession. I usually felt sick afterwards and then fell asleep.'

Penny, who is a 37-year-old public relations director from Suffolk, seemed only to crave sweet food during the 10 days each month before her period was due. Although she managed to restrict her intake to two indulgences per day, she found it worrying that her cravings got the better of her. No matter how resolutely she vowed that she would resist the cravings, she weakened each time they struck.

Another example is Jackie, a 36-year-old housewife, who although by no means a severe craver, felt the need to eat something sweet every day. Eating biscuits or chocolate daily made it impossible for her to lose any weight. In fact her weight crept up as the years passed, which she found immensely frustrating. Since following the WNAS programme, however, she has managed to overcome her sweet craving, and she has returned to her normal weight.

Severe cravings for food or drink can have disastrous consequences, as they may adversely affect your health, your behaviour and your self-esteem. The Sugar Craving Syndrome is often treated as a joke; rarely is it discussed in a serious manner. In fact, most sugar cravers do not even realise that there is a solution to their problem. As a result, they tend to go underground and to become secret bingers. I can still remember clearly my feeling of surprise when a patient first told me that her biggest daily problem was how to dispose of the chocolate wrappers so that her family would not find out about her habit!

June Mackie is a 34-year-old mother of three who works as a library assistant. She used to eat as much as two bags of 'fun-size' chocolate bars at one sitting:

'Chocolate was my biggest downfall. It was horrendous. I'd eat any kind of chocolate, including cooking chocolate if there was no alternative. I used to feel extremely guilty afterwards, especially when I had eaten all the children's sweets, particularly their Easter eggs or selection boxes at Christmas.

I was like a wild woman, capable of doing anything. I used to feel like I would die if I didn't have my "fix". I felt irritable and edgy. It's difficult to describe. Once I had eaten the chocolate I would feel like a great load had been lifted from my shoulders – and probably ended up round my hips! I then used to get dizzy and light-headed.

I followed the Nutritional Advisory Service programme and I noticed a difference within a month. I used to spend a fortune each month on chocolate. Now I am so much better. I have lost interest in chocolate and only eat it occasionally. I feel normal now, not like a demented person.'

Regardless of whether you simply need a daily 'fix', or your life is ruled by chocolate, there is a workable solution to hand. Over the years at the WNAS, we have had tremendous success with helping people to overcome their cravings for sweet food. The key to our success lies in our specialised nutritional programme which consists first and foremost of education on the subject of nutrient intake and specific dietary requirements that relate to each individual. In addition to this, we recommend daily exercise and nutritional supplements in the short term.

This book is divided into two parts. The first part is concerned with the important aspects of education and enlightenment on the subject of sugar and its effects on health; the second part deals specifically with a 30-day diet plan which is designed to help you overcome your cravings for sweet food. If you read Part I thoroughly, then follow the instructions in Part II carefully, within four weeks you will not only be well on the way to overcoming your sugar cravings, but will also be feeling generally fitter and healthier, and, as a result, will have a higher degree of self-esteem.

The first three chapters of Part I are full of interesting facts and figures. If you are desperate to get to the nitty-gritty straightaway, then you can read from chapter 4 onwards and come back to the facts and figures once you have made a start on the diet. Good Luck.

PART I
THE SUGAR CRAVING SYNDROME

1
Sugar the World Over

IN THE 1930s, the average Briton ate as many as four proper meals per day, with just one between-meals snack. Now, in the early 1990s, the average number of meals per day is down to two, with a staggering number of five snacks between meals. Chocolate is by far the most popular snack. In fact, in the UK, we spend more on chocolates and sweets than on tea, coffee and biscuits combined. We actually consume more chocolate than any other European country or the United States. Only the Swiss manage to polish off more chocolate than we do!

As well as being large chocolate consumers, the Swiss also enjoy large amounts of sucrose (table sugar), a fact that is paralleled by their high incidence of dental decay in children. The table overleaf lists, country by country, the largest consumers of both chocolate and sucrose-containing sweets.

Consumption of confectionery per head in 1990

	Sweets		Chocolates		Total	
	kg	lb	kg	lb	kg	lb
Switzerland	2.9	6.51	9.7	21.56	12.6	28.07
UK	5.0	11.24	7.3	16.28	12.4	27.52
Netherlands	5.6	12.50	6.7	14.96	12.3	27.46
Ireland	5.7	12.76	6.3	14.08	12.0	26.84
Germany	5.3	11.97	6.5	14.52	11.9	26.49
Belgium/						
Luxembourg	4.4	9.88	7.3	16.28	11.8	26.16
Denmark	5.1	11.35	5.6	12.54	10.7	23.89
USA	4.1	9.22	5.0	11.29	9.2	20.51
France	2.6	5.81	4.2	9.46	6.8	15.27
Italy	2.1	4.77	1.8	3.96	3.9	8.73

Annual review by Cadbury's and Trebor Bassett

Casting the Net Wider

The rate of increase in sugar consumption in developed countries over the last 200 years has been nothing short of phenomenal. At the end of the eighteenth century, annual consumption in the UK was about 2 kg (4½ lb) per head. It rose to a peak of some 50 kg (110 lb) per person per year in the 1970s, a rate of increase that matched that of population growth and the rate of monetary inflation. The whole of Europe, the United States, Australia and New Zealand have seen a similar marked growth in sugar consumption. Most of the increase occurred in the latter half of the last century and the first part of this century, when the manufacturing processes for extracting sugar from cane and beet were developed. Not surprisingly, those countries that were major cane sugar producers have had high levels of sugar consumption for long periods of time. This includes countries such as Cuba, Costa Rica, Fiji and Australia. Many African and Asian countries have extremely low levels of sugar consumption and with this low consumption, interestingly, goes a low incidence of dental decay.

In many Western countries, in recent years, there has been a trend towards purchasing and consuming less packet sugar. We bake fewer cakes and biscuits in the home today, and are more likely to purchase these ready-made from supermarkets. Many of us have given up sugar in tea and coffee, only to concentrate our attentions on confectionery, soft drinks and ice cream. The decline in our consumption of household sugar has been matched by a rise in the 'industrial' use of sugar. The three main 'industrial' uses are in chocolate and confectionery, biscuits and cakes, soft drinks and ice cream. Lesser amounts are used in the preserving of canned and frozen foods, jams and the brewing industry, while small amounts are used in pharmaceutical manufacture. As Mary Poppins so aptly pointed out, 'A spoonful of sugar helps the medicine go down'.

Turning up the Temperature

Interestingly, the warmer a country's climate, and the warmer the time of year, the less chocolate is consumed. The manufacturers have been mindful of this in recent years and, as a result, have invented the 'ice cream chocolate bar' in order to boost sales throughout the year. Once again, Britain has most of Europe licked when it comes to ice cream consumption. Our passion for ice cream far outweighs that of many of our Mediterranean neighbours. In fact, sales of ice cream chocolate bars in 1991 were more than double what they were in 1989. A report produced by Wall's in 1990 shows UK ice cream sales up by just over 62 per cent in five years. The table overleaf outlines European consumption of ice cream. Outside Europe, it seems that the Americans take the biscuit as far as ice cream is concerned, eating as much as 25 litres (5½ gallons) per person per year.

European ice cream consumption per head 1990

	litres	*pints*
Sweden	12.0	21
Denmark	7.9	14
UK	7.1	12½
Ireland	7.0	12¼
Switzerland	5.6	10
Netherlands	5.0	9
Belgium	5.0	9
Italy	5.0	9
Germany	4.8	8½
France	3.8	6½
Austria	3.8	6½
Greece	3.4	6
Portugal	2.8	5
Spain	2.5	4½

What is it about cold climates and sugar consumption? Chocolate and confectionery obviously will not withstand hot climates; indeed, for those who are physically active in a cold climate, sugar and sugar-rich foods provide a convenient and ready source of energy to help keep out the cold. In warmer climates, the greater availability of fresh fruits throughout the year provides a healthier alternative to sugar-rich foods.

A Ray of Hope

There have been some encouraging signs over the last decade. Many European countries have seen a small, 5–10 per cent, decline in total sucrose consumption. This is most often reflected in a fall in the domestic consumption of sucrose, while there is continued or even increased consumption of chocolate and confectionery, biscuits and soft drinks. Some of us at least are getting the message, particularly those who are well educated and have been exposed to the advice given by individuals and expert committees to moderate and reduce sucrose consumption.

Unfortunately, there is still an enormous surplus of beet sugar in Europe. This annual surplus has been estimated to be approximately 3 million tonnes. This is equivalent to approximately an extra 13 kg (30 lb) a year, for every man, woman and child in the EC. This 'sugar mountain' is not going to go away by itself. If the total consumption of sugar remains static, or falls slightly, in Europe, the obvious solution is to export surplus sugar to hungry Third World countries. This presents a moral dilemma. Sugar is certainly a cheap and sterile food, but it is devoid of all known essential nutrients. Getting the balance right for sucrose consumption in the Third World countries will require the judgement of Solomon to counterbalance the marketing expertise of Coca-Cola and other sweet food and drink manufacturers. The prospect is that European countries, who were net importers of cane sugar from the developing Third World countries 100 years ago, will become massive exporters of beet sugar to the developing Third World countries in Africa and Asia.

Hopefully, our nutritional experts will also export all we have learned about the relationship between sucrose and ill-health, especially dental decay. Only in this way will these countries be prevented from making the mistakes that we have made in Europe, the United States and other industrialized communities.

2

Sweet History

TABLE SUGAR, or sucrose, as it is more properly known, is a relatively new food; its presence and level of consumption have parelleled the technological advances of Western civilization. Sugar was first produced on a commercial scale when sugar-cane was planted and farmed in the West Indies in the eighteenth century. Prior to that, the most frequently used sweeteners worldwide were probably honey and some sugars derived from very sweet fruits, such as dates. With the Napoleonic wars, in the early part of the nineteenth century, trade between Europe and the West Indies was interrupted and sugar-beet began to be grown in northern Europe as a source of sucrose.

Sugar Production

Sugar-cane is a member of the grass family and grows like tall bamboo to a height of about 2½–3 metres (8–10 feet). Sugar-beet is a root crop and looks rather like a giant white parsnip. After harvesting, both crops are processed in a similar way. The leaves are removed, then the cane or beet is washed, crushed or cut, and put into tanks of heated water. The sucrose is dissolved out of the plant, residual debris is filtered away and the resulting liquid is then boiled to evaporate the water, leaving a thick syrup. The pure sucrose crystalises, leaving residual thick, dark syrups which form treacle and molasses. Nowadays, this process is highly mechanised and it is a testimony to modern farming and processing methods that sucrose is such a cheap form of calories. Without the technology of the modern world, such processing would not be possible.

Table sugar is a highly processed and pure food. It is

devoid of the fibre, vitamins and minerals that its parent plant contains in significant amounts. This loss of nutrients is a common result of processing. It also occurs, for example, in the refining of wholemeal flour to make white flour, when vitamin and mineral content is approximately halved. In the case of sucrose, however, the loss of nutrients is absolute; there are no vitamins or minerals in common table sugar.

What is Sugar?

If you wanted something sweet 200 years ago you would have had to rely on honey or the natural sugars that are present in fresh fruits and vegetables. Sugar comes in many different forms, the details of which border on the technical. Let us look, in simple terms, at exactly what sugars are and the different types that are available.

Sugars are sweet-tasting simple carbohydrates. 'Carbohydrate' is the name given to a varied group of foods which includes all the many different types of sugar, some more complex compounds and fibre. Carbohydrates form an important part of our diet and are found mainly in fruits, vegetables, grains, and related products. The average diet in industrialized countries provides about 50 per cent of calories as carbohydrates, while more primitive communities consume up to 80 per cent of their calories in the form of carbohydrates. Such a diet tends to be very high in fibre. Carbohydrates are usually a cheap and, at times, nutritious source of energy.

The different types of sugar are divided into two main groups: monosaccharides and disaccharides. More complex forms are called polysaccharides, a term which also includes starch and fibre; more of these later.

Monosaccharides

Monosaccharides are made up of one 'sugar unit'. They all have similar properties in that they form fine white crystals in the refining process and, of course, they all taste sweet. In nature, they are found dissolved in the water of the plant; hence fruit juices taste sweet. These single-unit monosaccharides include glucose (dextrose), fructose and galactose.

Glucose (dextrose) is found mainly in honey, some fruits, such as grapes, and in sugar substances derived from corn (corn syrup or corn sugar). Virtually all types of sugar in the diet are converted by the body into glucose which is used as the main source of energy by many body tissues, including those of the muscles, liver, kidney and brain. Brain and nervous tissue, in particular, is sensitive to changes in blood glucose levels. Although the term 'blood sugar' is often used, it actually refers to blood glucose.

Fructose is more commonly known as fruit sugar and is found in many fruits and honey. In fact, fructose is one and a half times sweeter than sucrose.

Galactose is a constituent of lactose, or milk sugar (see below).

During the process of digestion, carbohydrates are broken down into their component monosaccharides and are then absorbed. Some carbohydrates cannot be broken down by human digestion; these are known as fibre. Carbohydrates are digested very rapidly; most are digested within one or two hours of consumption, at which point the level of sugar in the blood rises to a peak.

Disaccharides

These sugars are made of two 'sugar units'. The group includes sucrose (table sugar) and lactose (milk sugar).

Sucrose (table sugar) is derived mainly from sugar-cane or sugar-beet. It is composed of one 'unit' of glucose and one 'unit' of fructose joined together. Sucrose occurs naturally in some fruits, notably dates and figs, and to a much lesser extent in other fruits, vegetables and grains.

Lactose (milk sugar) is composed of glucose and galactose (see above). It is the main type of sugar found in milk and related products and is much less sweet than sucrose. Some individuals, particularly of Oriental, Asian or Negro origin, have difficulty digesting lactose. Some Caucasian (white skin) individuals also experience difficulty, especially after a gastro-intestinal infection or acute diarrhoea episode.

As mentioned earlier, all sugars are converted into glucose, either during the digestive process or by the liver after digestion and absorption. The rise in blood glucose is obviously greatest after eating pure glucose. Fructose (fruit sugar) causes a very small rise in blood sugar, of only 5–10 per cent. As sucrose is composed of half glucose and half fructose, it causes approximately half the rise in blood glucose as that produced by the same amount of pure glucose alone.

Polysaccharides

These complex carbohydrates are made up of many 'sugar units' and include starches and fibre.

Starches are a type of polysaccharide that can be digested by the body. The commonest breakdown product of most starches is glucose. The rise in blood glucose levels after eating starch-rich foods, such as bread, pasta, rice and root vegetables, is variable. It is, of course, less than when eating the same weight of pure glucose.

Fibre consists partly of non-digestible polysaccharides, found in fruits and vegetables. There are different types of fibre with different properties. Some are particularly good at retaining water and others at retaining potential toxins or poisons that may be found even in the normal, healthy digestive tract. Some types of fibre, particularly fruit pectin, vegetables and oats, are considered to be capable of lowering an elevated blood cholesterol level, but fibre found in wheat bran is not effective in the same way. Fibre-rich foods also tend, in general, to delay the rate at which blood glucose levels rise after eating.

3
What the Experts Say

THERE IS OVERWHELMING EVIDENCE to support the fact that most of us should reduce the amount of sweet food we eat. Experts around the world agree with this; what remains a contentious issue, however, is precisely how much of a reduction should be made. Many expert committees in the United States, the UK, Australia, New Zealand and other countries have considered a wide variety of nutrition topics, including specifically examining the role of sugars in relation to disease. Surprisingly, their findings are not uniform.

For example, a report from the 1986 US Food and Drug Administration Sugars Task Force concluded that 'other than the contribution to dental caries [decay], there is no conclusive evidence that demonstrates a hazard to the general public when sugars are consumed at the levels that are now current and in the manner now practised'. These were carefully chosen words, as we will see later. In the United States, approximately 11 per cent of calorie intake comes from added sugars, mainly in the form of sucrose, fructose, corn syrup, which is high in fructose, and glucose, and 7–8 per cent from naturally occurring sugar. In the UK, the 1983 National Advisory Committee on Nutrition Education (NACNE) recommended a 50 per cent reduction in sucrose consumption, a view that was not accepted by many experts. The Department of Health's Committee on Medical Aspects of Food Policy (COMA) decided in 1986 that the role of sugar in the diet was worthy of a further study.

The COMA Report

In the UK, the Department of Health requested that the Committee on Medical Aspects of Food Policy (COMA) look at various aspects of health and nutrition. The COMA Panel on Dietary Sugars reported in 1989 in the publication *Dietary Sugars and Human Disease*. This COMA report stated that dental caries is positively related to the amount of non-milk extrinsic (added) sugars, such as sucrose, glucose and added fructose, in the diet and their frequency of consumption. This is undoubtedly true. The committee failed at that time, however, to make firm recommendations about sucrose consumption for the general population. The report stated that, in principle, the consumption of non-milk extrinsic sugars should be decreased and replaced with fresh fruit, vegetables and starchy foods. Although these foods are not entirely blameless with regard to dental caries, they do provide fibre and essential vitamins and minerals, and give a much better balance to the diet.

Fortunately, a later report, published by the COMA Panel on Dietary Reference Values, does give the practical recommendation of a 50 per cent reduction in the consumption of non-milk extrinsic sugars. This is not necessarily the ideal, but would represent an achievable goal that should yield substantial improvements in dental health and possibly an improvement in certain aspects of medical health as well. If everyone followed these recommendations it would help to reduce the incidence of obesity and, in the elderly, an increasing intake of non-refined carbohydrate sources, particularly vegetables and fruit, would improve overall nutrient intake and may help to reduce the risk of diseases such as heart disease and possibly even cancer. These effects, however, are likely to take many years or decades to become apparent.

Nutritional recommendations from the Canadian Department of Health and Welfare include minimizing consumption of refined sugars. Dietary guide-lines for

Australians also advise against eating too much sugar; similar advice has been given by the New Zealand Department of Health.

In order for the general public to consume less in the way of sucrose and added sugars, foods would have to be better labelled; it is estimated that as much as 80 per cent of the sugar we consume is now 'hidden' in sweet-tasting and other processed foods. The COMA report, *Dietary Sugars and Human Disease*, suggested that for those wishing to regulate sugar consumption, improved labelling of foods is essential. Manufacturers were urged to include the total sugar content of products on labels as well as, of course, details about fat, protein and fibre content. It was also recommended that the government seek the means for the analysing and labelling of all non-milk extrinsic sugars, thus enabling the monitoring of manufacturers' claims. These two steps would be warmly welcomed by health advisers who often feel that the good advice they give cannot easily be implemented by the public.

In the same report, COMA recommended that human nutrition should form an integral part of the training of medical students and other health professionals. This was a breath of fresh air to the team at the Women's Nutritional Advisory Service (WNAS) as we had discovered in a survey conducted in 1989 that 92 per cent of GPs admitted having no nutritional training. No wonder we have trouble trying to educate the public if the majority of doctors themselves are not actually trained in nutrition! If implemented, this would indeed be a breakthrough. Not only do doctors fail to learn about nutrition while training, but there is also little facility for knowledge about nutrition to be gained on post-graduate courses. The main reason for this seems to be that much of the post-graduate training is funded by the pharmaceutical industry, who simply have little vested interest in nutrition. The consequence of all this is that patients who read magazines and books and watch television are often better educated about nutrition in relation to their bodies than their doctors. To date, the government in the UK has shown little intention of funding education for doctors on the subject of nutrition.

A Question of Balance

To keep things in perspective, it is worth remembering that sucrose does have its good points. It is a very pure food and as such is safe; it is difficult for it to become a vector or agent for transmitting germs or toxins that may cause food poisoning or gastro-enteritis. The majority of us can consume large amounts of sugar in the short term without it having obvious or serious adverse effects on our health. Its benefits and its potential hazards are all a matter of balance.

Current levels of consumption for sugars (sucrose, glucose, fructose and honey) are 18 per cent of calorie intake in the United States and some 16 per cent of calorie intake in the UK. Approximately half of this intake comes from 'intrinsic' sugars that occur naturally in foods. The other half comes from 'extrinsic' sugars that are added to foods or beverages. The levels of consumption have been increasing over the last 150 years, although there has been some indication in the five years up to 1992 that consumption has reduced slightly in some industrialized countries.

Over 100 years ago, when sugar consumption was low, many disorders that are now relatively common were virtually unheard of. In particular, cardiovascular disease and diabetes were rare, if not almost unknown. Let us look briefly at what sugar might have to do with these and other diseases.

High-risk Groups

As we have seen, most experts agree that we eat too much sugar, and recommend reductions of about 50 per cent. These recommendations, however, are for the general public as a whole; for some specific groups a greater reduction is required to achieve optimum health. These groups include:

- the obese
- very high consumers of sucrose
- diabetics (insulin and non-insulin dependent)
- those with elevated blood fats (hypertriglyceridaemia)
- those who cannot tolerate sucrose
- those with reactive hypoglycaemia

Obesity

Some 25–30 per cent of the adult population of developed countries are significantly overweight. Obesity itself is associated with other conditions, such as diabetes, gall-stones, high blood pressure and increased risk of cardio-vascular disease. While there are undoubted genetic elements to obesity, in that overweight parents tend to have overweight children, the major contributing factor is the consumption of more calories than are needed to fulfil basic energy requirements. Excess calories, no matter what form they come in, are mostly converted by the body into fats, and are stored as such.

While you can put on weight by consuming too much of any kind of food, it is easy by comparison to consume excess calories by eating sweet foods. Sucrose-rich foods contain a lot of calories per mouthful, much more than the average mouthful of vegetables or fruit. All sensible weight-reducing diets recommend a reduction in high-calorie, nutrient-depleted foods such as sucrose-rich foods, alcohol and fatty foods. By cutting down on these foods, calorie intake is reduced without significant loss of essential nutrients such as vitamins, minerals, protein and fibre.

Diabetes

Insulin-dependent diabetics form approximately 0.5 per cent (one in 200) of the general population. Hopefully, all of these are on an appropriate diet and none needs to reduce added sugar consumption. Some 4 per cent of adults, particularly amongst the elderly and the obese, are non-insulin dependent diabetics. Up to 50 per cent of these may be undiagnosed and many of these, as well as some diagnosed non-insulin dependent diabetics, will have an added sugar intake that is too high.

Interestingly, the blood glucose control of non-insulin dependent diabetic patients may be improved with chromium, an important trace mineral which is found in many foods, notably complex carbohydrates, but is not present in sucrose (table sugar). A high intake of sucrose will lead to a reduced intake of chromium, and this may contribute to the diabetic state. Ironically, this important trace element is found in significant quantities in sugar-cane, as well as in many other plant foods. Mother Nature really did know best; the nutrients we need to metabolize our dietary sucrose are all available to us, provided we eat foods in their natural state!

Hypertriglyceridaemia

An elevated level of triglyceride fats in the blood affects only 1 per cent of the population, but again it is estimated that some 50 per cent of this group need to reduce their intake of refined carbohydrates. Losing weight and limiting intake of alcohol are also important in this condition.

Sucrose intolerance

An intolerance to sugar occurs in only 0.1 per cent (one in 1,000) or less of the population.

Reactive hypoglycaemia

This condition results when the body is unable to control blood sugar levels (see page 53). It probably affects less than 1 per cent of the normal population, but this is only a guestimate.

Figures published in the United States suggest that some 10 per cent of adults obtain 20 per cent or more of their calorie intake from added sugars, twice the national US average. Between 50 per cent and 70 per cent of these are not obese and would not therefore be identified as members of a high-risk group.

Effects on Mood and Behaviour

Sweet foods and drinks are often used as a 'pick-me-up'. 'Energy-giving' effects in this respect are mainly psychological.

It has sometimes been suggested that high sucrose consumption is a cause of hyperactive behaviour in children. Such claims seem to be more justifiable against colouring agents and food additives than against sucrose itself, though, of course, sucrose- and glucose-rich sweets and foods often contain large amounts of colouring agent to which an individual child might be sensitive.

Sucrose, in modest amounts, seems to be relatively harmless when consumed as part of a healthy, balanced meal. From personal experience, I would not want to deny my own four children the pleasant sensations associated with chocolate and sweets, but I do notice that their behaviour and concentration levels deteriorate if they eat sweet snacks on an empty stomach. In practice, it seems that children need a well-balanced, healthy diet and a sensible eating pattern, with sweet food eaten in moderation and preferably as part of or immediately after a meal.

The Verdict

Looking at it conservatively, some 35–40 per cent of the population *definitely* need to reduce sucrose consumption, even without considering associated dental problems. No wonder the US Food and Drug Administration Sugars Task Force had to choose its words so carefully in 1986. If these recommendations were followed to the letter, the loss of revenue to the sugar industry would be mind-boggling!

Qualities of sweet foods and sweeteners

	Dental caries	Western Diseases (e.g. cancer, heart disease)	Rise in blood sugar	Fibre content	Nutrient content
Sucrose	+++	+	++	—	—
Chocolate bar	+++	+	++	—	+
Glucose	+++	—	+++	—	—
Fruit	+	Protects against	—	++	++
Artificial sweeteners	—	—	—	—	—

4

The Ad-Man's Delight

IN THE AUTUMN of 1990, a major advertising campaign was launched by Silver Spoon and Tate & Lyle, manufacturers of beet- and cane-derived sugar in the UK. Their advertising campaign included well-casted television commercials (which were repeated in 1991) as well as double-spread advertising features in women's magazines. One particular magazine advertisement read 'Is sugar a natural part of our diet? Consult the experts'. The experts alluded to in this case were bees, birds and bears, who feed on nectar fruits and honey respectively. The advertisement continued: 'It's only humans who treat a "sweet tooth" as a bad habit. Wild creatures know better'. What the advertisers failed to mention was that these 'experts' also know better than to go to a supermarket to purchase packets of sugar, sticky buns, cakes, biscuits and vast quantities of chocolate bars!

The advertising campaign also, very rightly, urged us to consider 'Nature's rules for a healthy lifestyle. Balance. Variety. Moderation'. Wise words indeed, and more than we might expect from the advertising industry. However, 'A Mars a day' is not what I would describe as a dietary habit renowned for its balance and moderation; foods with added sucrose or glucose are rarely balanced. 'Variety' has become the watch-word of fast-food manufacturers rather than thought for the consumer. 'Moderation' would hardly be the best word to describe the aggressive and relentless advertising campaigns which are the trade mark of manufacturers of high-added-sugar foods.

Moreover, the same advertisement suggested that half of the 1,500 or so calories needed every day by the average person, should come from carbohydrates (sugars and starches). However, the advertisement failed to tell us what percentage of these calories should come from sugars and what from starches, or indeed what percentage their animal 'experts' take as sugar and starch. One piece of advice given, however, is highly commendable: 'If you like sugar, use it the way Nature does, to make nutritious foods delicious'. At last! Even Silver Spoon and Tate & Lyle together do not have the nerve to promote sugar consumption in soft drinks and sweets or fat-laden cakes, biscuits, buns and confectionery. Quite rightly, added sugars should be used to sweeten nutritious fruits, milk-based puddings and even some vegetables. In this way only can sugar manufacturers help nutritionists to encourage people in industrialized countries to achieve the dietary goals laid down by the real experts.

Unfortunately, most readers and viewers of such promotional advertising will not have the background knowledge and experience that would allow them fully to perceive the ambiguities in the persuasive arguments put forward. They are more likely to be tempted into a state of complacency or lulled into appeasement about their misgivings towards added sugar in the diet.

The Power of Advertising

From a report published by Keynote Publications in 1991, I discovered that the amount spent on media advertising of confectionery in 1990 in the UK was over £100 million. This money went into the advertising of chocolate bars, loose chocolates, boxed chocolates and sugar confectionery, including chews, jellies, pastilles, mints and gum. Big money is invested in 'building a brand', in other words in persuading us poor consumers that eating more of a particular chocolate bar, for example, would enhance our lives in some way. A classic example is the estimated £6 million that Cadbury's spent on launching their new 'Strollers' brand.

It is also interesting to note that 94 per cent of the advertising budget for chocolate and 96 per cent of the sweet advertising budget is used for television advertising. With this in mind, I decided to find out exactly what percentage of all advertising promoted chocolate, candy or sweet processed food. I enlisted the help of my brother-in-law, who is happy to use any excuse to watch television. He tuned in to a commercial television channel for a day and monitored the advertising. He discovered that 23 per cent of advertisements shown on that day promoted junk-food or chocolate bars. On hearing about such a tempting project, my children eagerly volunteered to watch great quantities of children's television in order to count the junk-food advertisements. We calculated that 41.6 per cent of the advertisements shown between children's programmes promoted chocolate, sweets or fast food, nearly double the amount of advertising time given to junk-food at other times of the day.

I also monitored the advertising at the cinema (on seven different occasions), again to determine what percentage of advertisements related to sweet foods or snacks. The vast majority of the advertisements were concerned with either alcohol, sweet processed foods or sweet snacks. Additionally, the cinema kiosks themselves were crammed full of junkie snacks and drinks; there were practically no nutritious foods or snacks on offer, so, to avoid temptation on your next visit to the cinema, it might be as well to take your own healthy snack along with you.

As you can well imagine, dental health professionals are up in arms about the number of advertisements that promote sweet foods and snacks, and about the claims that the advertisements themselves make. Early in 1991, the Mars company were at last asked to account for their famous slogan 'A Mars a day helps you work, rest and play'. And not before time. This slogan has existed for as long as I can remember, and I am sure that millions of people actually believe its powerful and persuasive message. Nutritionally speaking, it cannot possibly be true.

Many nutritionists, myself included, fail to be convinced by this exhortation that daily consumption of a Mars bar will improve work, rest and play. In the first place, in order to **work** efficiently it is best to eat foods rich in complex carbohydrates that break down over a relatively long period of time, thus keeping your brain well supplied with nutrients. A Mars bar is certainly not the best example of such a food! (On reflection, however, there may be some truth in their claim. Certainly consuming a Mars a day may mean a lot more work cleaning your teeth as well as seeing the dentist, and for some it may mean seeing the doctor, or attending a weight-loss clinic.)

If you dared to miss a meal altogether and replaced it with the 400 calories that a king-size Mars bar provides, an urgent need to **rest** may in the worst case be precipitated by reactive hypoglycaemia. As for **play**, this is certainly going to be needed if the 400 calories are surplus to your normal daily requirement. You would need to go for a five-mile run, play 30 minutes of squash, or make love three times every day in order to burn up the extra calories and avoid weight gain. On that note, I will leave you to decide how much faith *you* place in the claim that a Mars a day will help you work, rest and play!

Somewhat Misleading . . .

Another bone of contention is the fact that many advertisements for sweet snacks can mislead the public by omission. In other words, it is what the advertisement does *not* say that may be important. For example, the Department of Health's Committee on Medical Aspects of Food Policy (COMA) report in 1989 and the Food Advisory Committee report of 1987 both concluded that the word 'natural' could amount to a health claim. Experts agree that sugars that occur naturally in foods, 'intrinsic sugars', do not usually harm your health, while refined sugars certainly can. One recent advertisement in particular implied that processed sugar was 'natural' which gave a misleading 'healthy' image to the product in question. Official complaints were made to the Independent Broadcasting Authority (IBA) and the Advertising Standards Authority (ASA). The IBA rejected the complaints about the TV advertisements and the ASA stated that, in their view, the word 'natural' implied nothing more than the fact that the product was not derived from a synthetic source. It is indeed disappointing that the ASA seemed to be acting on advice from industry rather than on advice from expert committees like COMA.

It is likely that restricting advertising of chocolate and sweets would result in decreased sales. A model to support this theory is the Norwegian Tobacco Act, which came into force in 1975. This Act banned all advertising and promotion of tobacco products and strengthened health warnings about them. The outcome has been a steady fall in the popularity of smoking. On the other side of the coin, more recent Scottish research indicates that under-age cigarette smoking is increasing as a direct result of widespread advertising.

The relationship between sugar consumption and dental decay is now so thoroughly proven that dental health educators will continue their campaign until they achieve satisfactory results. The message is not to avoid sweet food altogether, but to eat it after meals and in moderation. As eating habits are developed at such an early age it is manifestly unfair that children should be bombarded with so many persuasive advertisements that encourage them to eat regular sweet snacks and consume endless cans of fizzy drinks, which may well establish a caffeine addiction, and will certainly lead to the development of a 'sweet tooth' and subsequent tooth decay.

5

The Sugar Craving Syndrome

IT IS PERFECTLY ACCEPTABLE to enjoy eating sweet foods from time to time and, as long as your diet is generally good, the odd indulgence should pass off unnoticed. After all, 'a little of what you fancy does you good'. The problem sets in when you cannot say no, when cravings get the better of you, and you find yourself driven out of the house after dark, looking for a late-night store that may supply your need! All the new year's resolutions in the world will not necessarily curb such cravings for sweet foods. However firm your resolve to leave them alone, when the 'munchies' appear your resistance is easily overwhelmed.

Why Do We Crave Sweet Foods?

There are often good reasons for these cravings; you can rest assured that you do not give in just because you are weak-willed. Cravings for sweet foods can be the body's way of saying that the blood glucose level is low. However, while a sugary snack may temporarily raise blood sugar levels, it is not the long-term answer, as the refined foods soon metabolize and the blood sugar levels fall again rapidly. Nothing short of changing your dietary habits and ensuring an adequate intake of vitamins and minerals will resolve the problem in the long term.

Sugar cravings might not be so difficult to come to terms with were it not for the fact that sweet foods taste so nice. The smooth texture is pleasing to the palate and the sweet flavour has long been associated with reward, possibly since childhood. It is all too easy for us to develop a psychological dependence on sweet food. We are continually being persuaded by aesthetic advertising that 'a Mars a day helps you work, rest and play' and that Coca-Cola is 'the real thing', not to mention the fact that 'a finger of fudge is just enough to give your kids a treat'. It all seems so ideal, but what the 'ad-man' doesn't tell us is that excessive sugar consumption is bad for mental and physical health, and that it rots children's teeth.

Ask yourself a key question: Does your eating pattern change according to how you are feeling emotionally? For example, if your love life is thriving, can you manage to maintain a healthy diet? But as soon as there is an upset, or things are not going so smoothly, do your dietary resolutions go out of the window? If you fit into this category, then it is more than likely that you are a comfort eater. Stress can also affect your eating habits; pressure of work or financial worries are just as likely to send you scurrying to the cookie jar or the corner shop. A poor diet, or negative emotions such as fear, depression, anger or anxiety, may affect you in a similar way.

What Can Be Done?

Instead of expecting you to change your lifestyle altogether, this book will show you how to look at your existing social and eating habits, and will help you to establish better habits, which, as a result, will improve your ability to deal with the problems in your life.

By changing your dietary habits, you will be able to keep your blood sugar on a more constant level and your physical and mental health will improve. As a result of feeling better in yourself, you will be able to deal more easily with the ups and downs of life, and bring about quick resolutions to problems. After all, if you're not feeling good, problems that occur tend to loom larger; once you feel 'on top', physically and mentally, you will find that life seems far more manageable.

Assessing the Damage

Let us now take a look at your eating habits to determine whether you are a high, medium or low sugar consumer. The questions in the following chart relate to your consumption of foods containing sugar over the last seven days. Before answering them, spend a minute thinking back over the last week and try to remember all the times you ate sweet foods. Now answer the questions, ring the numbers that apply to you, then total up your score.

Over the last seven days:	LOW (none/1)	MED. (2 or 3)	HIGH (4 or more)
How many chocolate bars or portions/servings of chocolate or sweets have you eaten?	1	2	3
How many portions of cakes, desserts or puddings, or 'portions' of biscuits (1 biscuit counts as ¼ portion) have you eaten?	1	2	3
How many cans or bottles of non-low-calorie soft drinks or servings of ice cream have you eaten?	1	2	3
How many portions have you eaten, either shop bought or home-made, of foods containing sugar, e.g. fruit pies, desserts, custard?	1	2	3
How many cups of tea, coffee, chocolate or other drinks *with sugar* did you consume per day?	1	2	3
TOTAL			

How did you score? A score of five or less means you are a low consumer. If you scored between six and 10, then you are a moderate consumer. A score of between 11 and 15 means you are a high consumer, which probably comes as no surprise to you!

The Sugar Craving Syndrome at Work

A street survey conducted by the Women's Nutritional Advisory Service (see page 38) revealed that, of the 500 women questioned:

- 60 per cent admitted that consuming sugar was a problem
- 72 per cent said they would like to consume less sugar
- 78 per cent wanted most of all to eat less chocolate
- 23 per cent said they became noticeably moody and irritable shortly after consuming sweet foods
- Just under 40 per cent admitted that weight gain was the worst problem created by their sugar cravings.

A more recent survey confirmed that chocolate was the most craved sweet food, and that weight gain was one of the biggest problems created. Weight gain and poor skin can both have a devastating effect on us emotionally, and on our morale. None of us like facing the world with excess baggage and looking spotty, do we? If you have gained weight as a result of your indulgences, take heart; you will find that the diet outlined in this book will help to normalize your metabolism so that your body has a chance to return gradually to its normal weight.

However, weight gain and poor skin are just two of a number of unpleasant symptoms that can result from an inadequate intake of nutrients allied to an excessive intake of sweet foods. Others might include the following:

- nervousness and anxiety
- fatigue

- headaches
- palpitations
- dizziness and fainting

Look at how these ghastly symptoms can affect our own lives and those of others around us:

Nicole Cohen is a 29-year-old single woman who works as a secretary. Her cravings for sweet food lasted for two weeks each month, during which time she would eat approximately one dozen chocolate bars per day:

'I used to crave chocolate, biscuits and cakes. I'd have to satisfy my urge instantly, and once I started I had to have more and more. I could never get enough sugar in those two weeks before a period.

I used to feel extremely irritable and agitated and ready to bite someone's head off. After eating the chocolate I would feel bloated, tired, like a complete blob. I used to spend at least £20 a month on chocolate before I followed the Advisory Service programme.

After just one month I felt I had completely overcome the cravings I had had for years. I am 100 per cent better, and am totally convinced about the whole programme. I did stray from the diet and noticed the problem returning; once back on the programme, however, the cravings abated again, this time for good.'

Another example is Elizabeth Marshall, a 28-year-old midwife. She had had weight problems since the age of 16, and severe sugar cravings which became worse 10 days before her period was due. On her worst days she would eat as many as six bars of chocolate, and she had cravings for junk food.

'I used to lay on the bed panic-stricken, wanting to die rather than face another 10 days of bingeing. My biggest nightmare was that the sweet shop would be shut and I would be unable to get my "fix". I used to feel on edge, "nervy", almost excited at the thought of eating lots of chocolate, dismissing any sensible thoughts as I just had to have it, whether it be first thing in the morning (before breakfast) or late at night.

After eating the chocolate I felt physically unwell. I had palpitations, sweating, bloatedness and lethargy. I felt guilty and loathed myself because I didn't have the will-power to abstain. I felt fat and desperate for my period to come so that the cravings would pass, vowing I would be better and try harder next month. I used to spend about £15 a month on chocolate.

I felt up-tight and irritable and would fly off the handle for no apparent reason. My marriage was under a tremendous strain.

I contacted the Nutritional Advisory Service for help. After being on their programme for a month, my cravings stopped.

Within two or three months, the problem had disappeared altogether, plus I had lost 9.5 kg (21 lb) in weight, which I thought was wonderful, especially as I wasn't dieting. I feel like I am back to my old self now; I'm energetic, enthusiastic and optimistic, and I feel that overcoming this problem has saved my marriage.'

6

Confessions of 'Chocoholics'

MY INVESTIGATION into sugar cravings began by accident some six years ago. The technical team at the Women's Nutritional Advisory Service (WNAS) undertook an analysis of 1,000 of our patients, all of whom had been suffering from Pre-menstrual Syndrome (PMS). It emerged that some 78 per cent of the women in the sample admitted that they craved sweet food, particularly before their period was due. We discovered that a large number of women with PMS were bingeing on sweet food to the point of making themselves feel utterly sick. They also reported extreme feelings of guilt and tremendous fatigue after their indulgence. As little research had been conducted in this area, we decided two years ago to undertake a further survey to determine whether craving for sweet food was a problem for women of all ages, and not just a problem for women suffering from PMS. The WNAS nurses conducted a national survey in the street; they interviewed 500 women of all ages and found that 72 per cent of the women surveyed would like to consume less sugar, and over 60 per cent of them admitted that craving for sugar was a problem for them.

Having determined that sugar cravings were common to women of all ages, we wanted to discover whether the same was true for men. On further investigation, we began to realise that many men also crave sweet foods, and give way to their cravings on a regular basis. The most commonly craved sweet food seems to be chocolate. The worst side effects reported as a result of bingeing are irritability, mood swings and fatigue. Weight gain is often a by-product of these cravings, although having said that, there is a substantial number of thin cravers, who manage to indulge without putting on weight.

Take the example of Graham, a 37-year-old who lives in Sussex. He has craved chocolate for as many years as he can remember:

'I would always break out in a sweat if I couldn't get my hands on a chocolate bar. On an average day I'd get through three or four chocolate bars and several Club biscuits, and wash them down with fizzy drinks. I also had cravings for chocolate digestives and chocolate nut biscuits. I could easily get through half a packet of biscuits in one sitting. My other weaknesses were for cake, particularly chocolate fudge cake, "death by chocolate" and chocolate mousse.

About half an hour after eating lots of chocolate I would feel happy and full of energy. This high feeling lasted for about an hour, after which I would then feel tired, lethargic and panicky. I was so dependent on my chocolate "fix" that I used to hide supplies of chocolate bars around the house, in case I had the cravings when the shops were shut.

I remember one occasion when I was doing a 100-mile time-trial on my bike. It was a time when I was consuming masses of chocolate and fizzy drinks, and eating very little good food. I was given a goody bag to begin with containing sugar lumps, a Mars bar and an orange. I ate the entire contents of the goody bag at the beginning of the trial. Halfway through I felt desperately hungry and dizzy. I had to stop and get something to eat. I staggered into a sweet shop and bought about eight chocolate bars and lots of drinks. I sat up on my bike and ate all the chocolate with no hands on the handle bars! I lost all the time I had made, and watched all the other riders overtaking me. I eventually finished the trial in 5 hours, whereas the previous year I had completed the same course in 4 hours 6 minutes. I coasted passed the finishing line, and almost fell off my bike on to the floor. I lay there feeling drained and like I wanted to die. Shortly afterwards I was very sick.

I used to think I needed glasses as my vision seemed to be disturbed and I had constant headaches. However, now that I have changed my diet, and am eating fruit and other foods with naturally occurring sugar instead of chocolate, I feel so different. I did have awful headaches while giving up the chocolate, but once I got through that I felt so much better generally. My eyesight seems fine now, and I hardly ever get headaches.'

Gluttons for Punishment

The most recent survey we undertook was a diet survey designed to determine which sweet foods were most commonly consumed, and in what precise quantity, in order to assess how much money these regular indulgences were costing per month, and what effect they were having on an individual's well-being. The survey consisted of a general diet questionnaire, included in which were several questions relating to the consumption of sweet foods and drinks.

Results

Following an appeal for sugar cravers in the newspaper, 295 women took part in the survey. The survey showed that 74 per cent of the women craved sweet food, particularly in the days before their period was due. The consumption of sweet food seemed to be related to age and occupation. Most of the women felt worse after eating sweet food, with fatigue being the most common complaint. The survey also revealed that people who consume lots of sweet food also tend to smoke more and drink more coffee, and eat less healthy foods such as fresh fruit, vegetables and salads.

We discovered that sugar consumption is related to occupation to a large degree. For example, retired women consume two and a half times more sugar than managers and directors. See how you score in the following table:

Total weekly sugar consumption by occupation

	teaspoons
Managers/directors	77.0
Skilled workers	100.1
Students	105.7
Unskilled/semi-skilled workers	110.6
Housewives/unemployed	123.6
Retired women	192.6

Sugar consumption seems also to be related to age and weight in that the older, larger women in the group ate the most sweet foods.

When asked which foods they most commonly craved, a vast majority of the women in our survey included chocolate in their list, with cakes coming a close second.

Foods most commonly craved

Food	No. of women (per cent)
Chocolate	91
Cakes	87
Puddings	66
Honey/jam	54.6
Ice cream	50.9
Biscuits	50
Sweets	40
Soft drinks	28
Other foods	25

We also asked our sample of women to estimate how many times each week they usually ate these foods. Again, chocolate, cakes and biscuits proved the most popular.

Weekly consumption of sweet foods

Food	Portions per week
Chocolate	4–7
Cakes	4–7
Biscuits	4–7
Soft drinks	1–3
Ice cream	1–3
Puddings	1–3
Honey/jam	1–3

Interestingly, we also discovered that the number of cakes, biscuits, ice cream, puddings, honey and jam consumed tended to increase with age, while consumption of chocolate, soft drinks and alcohol tended to decrease with age.

The cost of the cravings

The last thing a 'chocoholic' usually does is to stop to work out just how much money he or she is spending on chocolates and other sweet foods. We were able to calculate that the average amount of money spent by each person in the survey per week on sweet food was £5.57.

Weekly cost of cravings

No. of women (per cent)	Amount £
23	1–2
44	4–6
15	7–9
6.7	10–12
1.3	13–15

In our survey, 9.4 per cent of the women said they did not know how much they spent on sweet food!

Well-being in relation to consumption of sweet food

Physical and mental well-being of the women in our sample seemed to be seriously affected by the consumption of sweet foods. The diet survey showed that after eating sweet foods, 69.1 per cent of the women actually felt worse; 16.6 per cent felt better and 13 per cent felt no different. The women who felt worse after indulging experienced the following symptoms:

Effect on well-being

No. of women (per cent)	Symptom
54	fatigue/tiredness
40	depression
35	sickness
33.9	headache
20	anxiety
6	shakes

On a Lighter Note

You may be surprised to learn some of the following facts:

- Over the Christmas period, shoppers snap up £400 million worth of selection boxes.
- 300 million Creme Eggs are made every year.
- In the 1600s, some people believed that women who drank too much chocolate would give birth to darker-skinned babies.
- The Duchess of York named her dog Bendicks, after the makers of her favourite chocolate treat.
- Women buy 65 per cent of all confectionery, but eat only 39 per cent. Men do 27 per cent of the buying and eat 26 per cent of it themselves. The other 8 per cent – children under 14 – do 35 per cent of the eating, so they're bought a lot by their mums.
- If all the Crunchies sold in a year were placed end to end they would stretch from here to Bangkok and back.
- Money spent on confectionery each year in Britain would buy 5,000 Rolls Royces.

7

Sweet Nothing – The Nutritional Value of Sugar

SUGAR is undoubtedly an attractive food. What makes it so is not its nutritional value but its other qualities. In particular, its sweet taste is considered by virtually all populations and individuals to be desirable; most naturally sweet-tasting foods, such as some fruits, honey and some vegetables, are sought-after commodities. On the other hand, bitterness is not a desirable taste and is sometimes associated with poison. Traditionally, sweet-tasting foods are · viewed as being desirable and beneficial, and bitter-tasting foods as being undesirable, potentially harmful even. Sugar has the ability to mask bitterness, and to counteract the acidity found in many sharp-tasting fruits, such as lemons, grapefruit and some soft fruits.

A historical exception to this rule is wine 'sweetened' with lead. In ancient Rome, rough-tasting acidic wine, which was rich in acetic acid, was placed in lead casks to 'sweeten' it. The acetic acid would combine with the lead of the container to form lead acetate. Surprisingly, this has a sweet taste and is hence known as 'sugar of lead'. Indeed, it forms crystals which appear very similar to sugar itself. Significant, even toxic, amounts of lead could be found in wine sweetened this way. It was felt that lead contamination of both the wine and the water in ancient Rome contributed to the downfall of its empire.

In Sugar's Favour

Besides sweetening foods, sugar has other useful culinary properties. It provides bulk which, for example, is useful in the making of meringue; it can be used to form syrups which have attractive 'mouth feel'; it is a useful preservative as in high concentrations it impedes the growth of bacteria and moulds; and it can be caramelised, that is it can be heated until it turns brown, then used to impart both colour and flavour to foods. There is little doubt that sugar is a versatile and desirable culinary commodity.

On the Down Side

As a basic, staple food, however, sugar has little to commend it other than its high number of calories. Sucrose, glucose and fructose are pure foods and contain no fat, fibre, vitamins or minerals. The absence of fat makes it useful as a food only for those few individuals who are intolerant of fat yet require plenty of calories.

The two most convincing nutritional arguments against sucrose and its relatives are that they are not balanced foods in themselves, providing none of the essential vitamins and minerals our diet requires (let alone those necessary for the metabolism of glucose), and that in industrialized societies, sugars are used to make palatable foods that are high in fat and salt and low in fibre and essential nutrients. While sucrose itself may be of limited harm, it leads us into temptation and off the nutritional straight and narrow.

The Effect of Excessive Consumption

If you satisfy your calorie requirement with sweet food, it is likely that you will consume less of the more nutritious and well balanced foods. If added sugars compose only 5 per cent of your calorie intake, a level in line with certain expert recommendations, the potential loss of essential nutrients is relatively small. However, some 10 per cent of the population consume 20 per cent of their calories in the form of added sugars. This may not matter if you are physically very active and the rest of your diet is very well balanced, but for those who are not physically active and have an average calorie requirement, including, for example, many young to middle-aged women, and some elderly people, this high consumption of sucrose reduces their potential intake of essential vitamins and minerals by 15 per cent.

The effect of excessive consumption becomes particularly important when nutritional demands are greatest, such as during times of physical ill-health, mental ill-health or stress, or when pregnant or breast-feeding. Children with small appetites are also at risk. Workers from the Nuffield Laboratories of Comparative Medicine at the Institute of Zoology in London looked at the dietary intake of 419 pregnant women. Calculation of their food intake showed that 25 per cent of women with the highest added sugar intake had vitamin and mineral intakes of between 20 and 50 per cent lower than the 25 per cent of women with the lowest intake. Nutrients such as vitamin B_1, vitamin B_3 and magnesium, which are needed to metabolize the increased amount of carbohydrate, were substantially reduced.

A report from the US Food and Drug Administration Sugars Task Force expressed the view that 'there is no firm evidence which supports the contention that sugars, as they are currently consumed within the USA, interfere with trace minerals' metabolism and contribute to mineral deficiencies'. Not everyone, including myself, agrees. In fact, the health education authorities in the UK, on reviewing the report of the panel on dietary sugars from the Department of Health's Committee on Medical Aspects of Food Policy (COMA), concluded that 'people who eat only a small amount [of food], choosing dietary sugars rather than more nutrient-dense foods, may limit the intake of other nutrients. For such people, appropriate selection of food items is of special importance'. Such advice may apply in particular to about 10 per cent of the healthy population, and to a higher percentage of those who have health problems.

The Nutrient Content of Sweet Foods

The chart on page 47 lists the sugar, fat, fibre and calorie content of various sugar-rich foods, including some fruits, vegetables and confectionery. As you can see, most sugar-rich foods are low in fibre and some, such as chocolate bars, biscuits and cakes, are high in fat. Their naturally sweet counterparts, such as grapes, figs and carrots, on the other hand, are low in fat and high in fibre, as well as having a lower calorie content per mouthful. Sugar-cane and sugar-beet, from which our sugar is derived, are similarly low in fat and high in fibre. You can see from the chart just how low the nutrient content of sugar and some of the major sugar-rich foods is.

Naturally sweet foods, including sugar-cane, sugar-beet, figs, grapes, carrots and apples, provide significant amounts of vitamin C, vitamin B and magnesium, as well as other minerals and fibre. Chocolate bars, and cola-based drinks are extremely low or completely devoid of such nutrients. Jam, stewed fruit with sugar and ice cream occupy an intermediate position. A high-sugar processed food, such as McDonalds' apple pie, is a typical example of a high-sugar, high-sodium (salt), low-fibre, nutrient-depleted 'food'. Nutritionally speaking, a McDonalds' apple pie is about as well balanced as an overweight tightrope walker!

Nutritional content of sugar-rich foods per 100 g (4oz)

Food	Calories	Fat	Fibre	Sugars
Raw carrots	23	trace	2.9 g	5.4 g
Frozen peas, boiled	80	0.4 g	12.0 g	1.0 g
Eating apples	46	trace	2.0 g	11.8 g
Bananas	79	0.3 g	3.4 g	16.2 g
Mars bar	441	18.9 g	none	65.8 g
Bounty bar	473	26.1 g	none	53.7 g
Toffee	430	17.2 g	none	70.1 g
Bread				
White	233	1.7 g	2.7 g	1.8 g
Wholemeal	216	2.7 g	8.5 g	2.1 g
Biscuits				
Plain digestive	471	20.5 g	3.1 g	43.6 g
Chocolate-coated				
digestive	524	27.6 g	5.5 g	16.4 g

Should Sugar be Excluded from the Diet?

If they are of such limited nutritional value, should sucrose and other added sugars be excluded from the diet altogether? Of course not. The best way to use sucrose and other sugars is to use them in moderation in a way that encourages the intake of nutritious low-fat, high-fibre foods. There are many ways in which to do this, for example stewed fresh fruits need a little sugar to make them palatable, and many nutritious desserts, such as apple pie or fruit salad, are improved by the addition of a little sugar or sugar syrup. Desserts made with nuts and fruit are particularly nutritious and may require sweetening; you will find recipes for some of these desserts later in the book (see pages 105–181). Low-sugar, high-fruit jams contain modest amounts of fibre and vitamin C; spread on wholemeal bread they make an appetising and nutritionally well balanced snack.

Milk puddings, such as rice pudding, blancmange and chocolate mousse, are a useful way to provide calcium for children (and grown-ups!). How many children would eat them if they didn't contain a *little* sugar? The same is true for some ice creams, which may be made yet more nutritious by the addition of fresh fruit or nuts. Caramel custard employs sugar both as a sweetening agent and to give the custard a delicious caramel colour. Sugar even enhances the flavour of vegetables; the ever-popular and convenient can of baked beans is surprisingly nutritious, being rich in fibre, protein and several vitamins and minerals. The tomato sauce is, of course, sweetened with sugar, although reduced-sugar and sugar-free versions have recently become available. One other way in which sweet foods can encourage an intake of other nutritious foods, is if they are given to children as a reward for a clean plate, though it is important that the nutrient-rich foods are eaten first; children should not be allowed to fill up on sweet foods.

If used in moderation in this way, added sugar has a useful role to play in the diet. Part II of this book will teach you how to use naturally occurring sugars and added sugars in your diet in a nutritious way, while helping you to break the craving habit.

8

All About Low Blood Sugar

WE HAVE PROBABLY all experienced mild symptoms suggestive of reactive hypoglycaemia, or low blood sugar. These might occur if we have missed a meal or, in particular, if we have had an alcoholic drink instead of a meal or an hour or two before eating. At such times there is a tendency for blood sugar, or blood glucose, levels to dip, and with them our energy levels. As hunger is one of the symptoms of hypoglycaemia, craving for a quick snack, especially an energy-giving one, can understandably be the result. Ingestion of a sweet, sugary snack, particularly one rich in sucrose or glucose, causes a rapid increase in blood sugar which will help to satisfy not only the hunger but also the symptoms of hypoglycaemia, but only in the short term.

The wide availability of sucrose snacks makes it easy for us to turn to them as a quick solution to our problem. Once we have 'learned' that a biscuit, chocolate bar or piece of cake quickly satisfies that hungry and faint feeling, a view that is reinforced by manufacturers' advertisements that inform us of their products' energy-giving properties, a pattern for craving for such foods can be established. Thus, a combination of physical, psychological and social factors can easily lead to sugar and carbohydrate cravings. It is far better, in the long term, to consume wholesome nutritious foods as a means of maintaining blood glucose levels, rather than resorting to refined processed foods when cravings strike.

The main sugar in the blood is glucose, which can loosely be regarded as a universal fuel for most body cells.

It is usually the premier fuel used by muscles, and indeed the only form of energy that can be utilized by the brain, spinal cord and nerves. The glucose in the blood is derived from two main sources: Firstly, the carbohydrates that we eat are broken down into their component single sugars (monosaccharides), the commonest of which is glucose (see page 18). Other single sugars, such as fructose and galactose, are converted, albeit slowly, into glucose. A carbohydrate-rich meal causes a rise in blood glucose levels. The rapidity and degree of rise varies considerably, depending upon the source and quantity of carbohydrate eaten. If your last meal was the only source of glucose in the blood then, of course, this would fall to nothing with fasting. Blood glucose has, however, to be maintained within clearly defined lower and upper limits for good health.

How Blood Sugar is Controlled

Blood glucose levels are prevented from falling too low by the production and release of glucose from the liver. The liver is able to release glucose from its store of glycogen, which is usually depleted within 48 hours of absolute fasting. More glucose, however, can be formed from fats, proteins or other sugars, as necessary. The liver is thus the key organ for controlling blood glucose levels and, in particular, for preventing an excess fall. Several important hormones also influence blood glucose levels.

Maintaining the balance

You have probably all heard of insulin. This hormone is produced by the 'beta-cells' of the pancreas and helps lower blood glucose levels by stimulating the passage of glucose from the blood into other body cells. Many diabetics require injections of insulin as their own pancreatic cells which normally produce it have been destroyed or cannot respond to a rising blood glucose level. Thus, in diabetes, blood glucose levels are high. The excess blood glucose can spill into the urine and the presence of sugar in the urine is one way in which diabetes is classically diagnosed.

Several other hormones are responsible for helping to raise a low blood glucose level. They include glucagon, adrenalin (also known as epinephirne) and corticosteroids (commonly known as 'steroids').

Glucagon is also produced by the pancreas but from different cells called 'alpha-cells'. It causes a rise in blood glucose by stimulating the production of glucose in the liver and its release into the circulation. Like insulin, glucagon is extremely powerful and very quick-acting. It is secreted when glucose supplies are running out, such as during starvation, after physical exercise and when the blood glucose level is low (hypoglycaemia).

Adrenalin (epinephirne) is produced by the two adrenal glands adjacent to the kidneys; its purpose is to stimulate a rise in blood glucose. Adrenalin is the hormone produced in response to fight, fright or flight and can loosely be termed the 'hormone of stress' as it prepares the body and its metabolism so that it can cope with sudden physical activity or stress. It stimulates the release of glucose from stores both in the liver and the muscles in readiness for its utilization during increased physical activity. Unlike glucagon and steroids, however, adrenalin has some other important effects which contribute significantly to the symptoms of hypoglycaemia.

Corticosteroids are another group of hormones produced by the adrenal glands. Some of them have the effect of raising blood glucose levels. Cortisone and other steroids, which are used medically in the treatment of arthritis, skin disorders and many other chronic inflammatory diseases, have the unwanted side-effect of raising the level of blood glucose and inducing a diabetic-like state.

Hypoglycaemia

The complex system that controls blood sugar levels can sometimes break down, allowing blood sugar levels to fall dramatically as glucose is used as an energy source by the body's vital organs, including the muscles and the brain. The disadvantage of this is that the tissue of the brain and nervous system depends on glucose, and a severe fall may cause profound changes in mental function, physical ability and mood. Weakness, fatigue, poor concentration, hunger, a fall in blood pressure and a pale appearance are common symptoms caused by a reduction in the supply of glucose to the brain and nervous tissue. The release of glucagon itself helps to correct this, as does adrenalin, but it can be several minutes, up to 1 hour, before full recovery is experienced. Glucagon has no additional effects, but adrenalin has a powerful stimulant effect, causing a rise in pulse-rate, blood pressure, sweating, palpitations and a feeling of anxiety and panic. Severe hypoglycaemia, if sustained, can cause changes in the electrical activity of the brain and even epilepsy in susceptible people.

Symptoms of Hypoglycaemia (low blood sugar)

Due to lack of glucose (energy) for the brain

Weakness
Faintness
Irritability
Hunger
Nausea
Headaches (migraine)
Dizziness
Fatigue

Due to adrenalin stimulation

Palpitation (rapid heart beat)
Sweating
Pallor
Anxiety
Apprehension

There are three main types of hypoglycaemia. The most common type is that caused when an insulin-dependent diabetic either injects too much insulin or fails to eat soon enough, or to eat an adequate amount of food, after an insulin injection. The effect of an excess of insulin is to lower the blood glucose to levels that can cause a marked reaction.

Hypoglycaemia can also occur as a result of certain diseases. Very rarely, a tumour of the pancreas may secrete insulin and cause profound hypoglycaemia, particularly if the person goes without food for several hours. Hypoglycaemia can also accompany disorders such as liver disease, especially alcoholic liver disease, over- or under-activity of the thyroid gland, as a side-effect of some drugs, after operations that change the emptying pattern of the stomach and after alcoholic binges.

Reactive Hypoglycaemia

Reactive hypoglycaemia is a term used to describe a fall in blood sugar following a rise brought about by eating a meal with a high content of starch or refined carbohydrates, sucrose or glucose. There has been much debate over the significance of reactive hypoglycaemia. Early work with many types of patients, particularly those with symptoms of anxiety or fatigue, suggested that excessive dips in blood glucose levels after a meal were quite common. More careful assessment has, however, revealed that blood glucose has to fall to very low levels before it will produce definite symptoms.

True severe reactive hypoglycaemia is probably much less common than was at first thought. It is more probable that a drop in blood glucose towards the lower end of the normal level is also accompanied by normal feelings of hunger as well as other metabolic changes that signal to us that it is time to eat. The exact nature of these chemical signals has not yet been worked out but they probably involve changes in carbohydrate and protein metabolism in the brain.

The tendency to mild or indeed significant reactive hypoglycaemia can be worsened by the following factors:

Obesity is associated with poorer blood sugar control with a tendency to both high and sometimes low dips in blood sugar. At its worst, this can sometimes precede the diabetic state.

Alcohol can cause severe liver damage which can lead to significant hypoglycaemia. Replacing your normal lunch with two or three lunch-time gin and tonics can cause a profound rise and subsequent fall in blood glucose levels, producing all the symptoms of hypoglycaemia. The natural rise and fall of blood glucose levels that occurs after consuming carbohydrates is enhanced by alcohol.

Smoking can increase the release of insulin and also glucagon; a further factor in destabilizing the smooth control of blood glucose levels. Smoking is also a powerful appetite suppressant in the short term. If a balanced meal is not consumed, blood glucose levels will inevitably fall.

Tea and coffee, if consumed in large amounts, can also cause an increase in the release of insulin. Large amounts of sugar consumed in tea or coffee can also contribute to an unstable blood glucose level.

Eating irregularly can also lead to swings in blood glucose levels. This is particularly significant for insulin-dependent diabetics. Missing a meal or fasting will cause blood glucose levels to fall to the lower end of the normal range in susceptible individuals.

Interestingly, exercise is the one factor that can improve the control of blood sugar, as well as having many other health benefits. Exercise increases the sensitivity of the body's response to insulin, leading to smoother control of blood sugar levels.

How Different Food Groups Influence Blood Sugar Levels

The rise in blood glucose levels after eating certain foods can be expressed by a measurement called the 'glyceamic index'. Obviously, the highest rise in blood glucose will occur after eating glucose itself. The degree of this rise, both in quantity and duration, is measured. The degree of rise in blood glucose levels after eating an equivalent amount of calories in other types of food is measured and expressed as a percentage of the rise caused by pure glucose. Fructose (fruit sugar), which contains no glucose and is only slowly converted to it by the liver, causes only a fraction of the rise, 10 per cent of that produced by glucose. Sucrose, which is 50 per cent glucose and 50 per cent

EFFECT OF DIFFERENT FOODS ON BLOOD GLUCOSE LEVELS

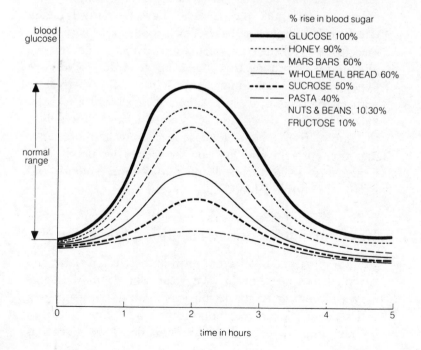

fructose, predictably causes 50 per cent of the rise produced by the equivalent weight of pure glucose.

The chart on page 55 clearly illustrates the different rises in blood glucose level caused by different types of food. It may surprise you to see that honey causes almost as high a rise as glucose itself. Mashed potatoes, white bread, rice and new potatoes all cause a substantial rise in blood glucose levels of between 70 per cent and 90 per cent. Wholemeal bread and Mars bars seem to be no different at 60 per cent! This is because the starch of wholemeal bread is broken down so rapidly that its glucose content is released just as quickly as that from a Mars bar. However, it must be remembered that wholemeal bread and other whole grain products are much more nutritious than chocolate bars, and that they will provide the vitamins, minerals and fibre necessary for a healthy metabolism. In the short term, there may be no difference as far as blood sugar control is concerned, but in the long term, there may be a big difference in overall health.

Peas, beans, nuts and seeds all have a relatively small effect on blood glucose levels. Such foods are now considered to be excellent components of a diabetic diet and are also useful in helping to control any tendency to reactive hypoglycaemia. Combining any source of carbohydrate with fat or protein, as in a balanced meal, will also help. Fat and protein delay stomach emptying and thus reduce the rate of rise in blood glucose levels. Furthermore, after digestion, protein and fat can themselves be slowly converted into glucose, thus leading to a more smooth and balanced control of blood sugar levels.

Key Nutrients Influencing Blood Sugar Levels

Certain nutrients also have a profound effect on sugar and carbohydrate metabolism. We treat glucose in the same way a motor car treats petrol; it is literally burnt up by a series of interlinked metabolic steps which convert carbohydrate into water and carbon dioxide. This process is called 'The Kreb's Cycle'; it was discovered by a famous German-born biochemist, Hans Kreb, who, together with Fritz Lipmann, was awarded the Nobel Prize for his discovery. The Kreb's Cycle requires various vitamins and minerals for it to function effectively. The vitamins required include vitamin B_1 (thiamin), vitamin B_3 (niacin) and vitamin C. Glucose is first broken down into shorter fragments which then enter The Kreb's Cycle, releasing energy in a series of stages. The residual products of these reactions combine with more glucose fragments to maintain the cyclical energy-releasing process.

Other nutrients are also important to the control of blood glucose. Insulin, for example, requires the trace element zinc in order for it to have the right shape and structure. As we have seen (see page 25), the trace element chromium appears to be of particular importance, especially for the elderly and some non-insulin dependent diabetics. It seems that chromium improves the response of tissues to insulin. In this way, chromium may help to stop excessive swings, both rises and falls, in blood sugar levels. Chromium Glucose Tolerance Factor (GTF), an organic chromium compound, is particularly useful in helping to control blood sugar levels; it is more biologically active than other forms of chromium.

Magnesium is another mineral that has been shown to influence blood glucose control. One study, by Dr Jo-Ann Kelly and colleagues, in the United States, involved giving a magnesium supplement to eight women. The rise in blood glucose after eating a standard amount of glucose was reduced by magnesium supplementation.

One of the major disadvantages of the processed foods found in industrialized societies today is the reduction in their nutrient content when compared with the original raw ingredients. Magnesium, chromium, other minerals, and vitamins B and C in particular are easily lost in the refining process, as we have already seen. Their loss from the diet increases the risk of even mild deficiencies occurring in susceptible individuals.

The Women's Nutritional Advisory Service (WNAS) has found that supplementing the diet with these vitamins and minerals has proved particularly useful in helping to control blood sugar levels. In the early days, we recommended that patients took extra supplements of chromium GTF, magnesium, B vitamins and vitamin C. While this cocktail proved effective in many cases, it involved taking lots of different pills which worked out expensive and cumbersome. Eventually we formulated a single supplement, called 'Sugar Factor', which contained all these nutrients at their optimum levels to take the place of the cocktail. We have been using this supplement successfully for the last four years.

Sugar Factor has recently been the subject of a clinical trial in which 100 women participated. The women were divided into two groups, with 50 in each. One group took Sugar Factor and followed the WNAS dietary recommendations, while the other group took dummy tablets and made insignificant changes to their diet. The group of women taking Sugar Factor experienced greater relief from their sugar cravings. Fifty per cent of this group overcame their sugar cravings, and a further 28 per cent reported some modest improvement, whereas only 36 per cent of those taking the dummy pills achieved the same result. Although further research needs to be carried out, we were pleased with these initial results.

The diet in Part II of this book is designed to help you overcome excessive dependence upon sugar and other refined carbohydrates by replacing them with nutrient-rich foods that will help to maintain optimum blood sugar control. Concentrate on eating a diet high in chromium, magnesium and vitamins B and C, as detailed in the tables on pages 98–104. Additionally, if you suffer with substantial sugar cravings, you may like to combine the diet with a course of Sugar Factor (see page 184 for availability), at least for the first few months, to help control your blood sugar levels.

9

A Sweet Tooth

SUGAR DOES NOT only affect mood, behaviour and body size; it can also have a disastrous effect on dental health. There is a substantial relationship between diet and tooth decay, or dental caries as it is correctly known. Tooth decay is the second most prevalent disease of mankind, after the common cold. The incidence and severity of this disease has climbed dramatically since the Industrial Revolution. Fortunately, preventative measures implemented over the past 25 years have started to reverse this historic trend. A new generation is now reaching adulthood with more teeth and less fillings. However, their parents, in the 30–50 age bracket, were not so fortunate and often have heavily restored teeth.

The 1989 report from the Department of Health's Committee on Medical Aspects of Food Policy (COMA) reported that dental caries was still very prevalent in the UK and that the decrease in the incidence of caries over the last 15–20 years appears to have slowed considerably. In 1973, the first UK survey of child dental health showed that 71 per cent of 5-year-olds and 97 per cent of 15-year-olds had experience of dental decay. In 1983, in the UK, virtually half, 49 per cent, of 5-year-old children had one or more decayed teeth and 93 per cent of 15-year-olds were affected. More recent reports show that by 1987 the incidence of dental decay in 5-year-old children was down to 45 per cent. Levels of dental decay are at last declining in the United States, but the Food and Drug Administration still admits that the present levels of sugar consumption within the United States contribute significantly to dental caries. A similar situation exists in Canada, Australia and New Zealand. The decline in dental decay has been attributed to the use of fluoride in toothpaste and in drinking water, rather than to any changes in sucrose consumption.

Over the years, much has been written by the dental profession on the subject of sugar and disease. Early and respected professionals, like Professor John Yudkin, realised quite some time ago that there was a link between sugar and disease and in fact wrote the popular scientific book about sugar called *Pure, White and Deadly* (see page 183).

What are the Causes of Tooth Decay?

Diet is not the only factor that relates to tooth decay. In order for tooth decay to occur, bacterial plaque must be present in the mouth. Plaque is the off-white, creamy deposit that begins to collect on your teeth even within an hour of brushing them. If you have not previously been aware of plaque, try running your tongue around your teeth just before brushing them in the evening. Plaque feels like a rough film to your tongue.

The mouth is an excellent environment for bacterial growth because of its humidity, its temperature and its rich supply of nutrients from the diet and saliva. Although dental plaque contains dozens of different bacteria, there is one particular offender, *Streptococcus mutans*, that seems to play the biggest part in causing tooth decay. The bacteria have a prodigious ability to produce acid from fermentable carbohydrate (sugars), and are able to survive in a highly acidic environment. They are also capable of producing a sticky substance from sucrose that increases its ability to accumulate and stick on to the teeth.

When you eat a sweet, sugary snack, the bacteria in the plaque on your teeth ferment the residues of food or drink that are left in your mouth. The result of this fermentation process is that an organic acid is produced which then wears away at the surface enamel of the teeth. If this process is allowed to persist, after about three weeks a small white spot will appear on the tooth, which is the initial site of demineralisation. It is not until some time later that you would feel pain or actually notice a hole in your tooth that needed filling.

By improving the standard of your dental hygiene, in other words by removing the bacterial plaque regularly (at least twice a day), and by adjusting your diet, you can go a long way towards avoiding dental decay. In addition to these two measures, you can use a fluoride-containing toothpaste which will help to protect the teeth from decay. In fact, in areas where fluoride occurs naturally in the water, the incidence of tooth decay in children is much lower.

The relationship between diet and tooth decay

The relationship between diet and tooth decay has been suspected for centuries, but it wasn't until 1950 that studies proved that decay only occurred when a diet containing sugar came into contact with the teeth. We now know that the common sugars, such as sucrose, lactose, glucose, fructose and maltose, present in fruit, dairy products and other sugar-containing foods, are all readily fermentable. It is, therefore, extremely important, from a dental health point of view, to restrict the intake of sweet snacks to a maximum of one per day, after which, ideally, you should clean your teeth, or at least swish some water around in your mouth to dilute the sugar after eating.

The most recent evidence suggests that sweet food is relatively harmless to the teeth if eaten only at meal times, that is no more than three times a day. This confirms other researchers' findings that sweet foods consumed as snacks between meals will cause harm.

Beware of the healthy snack

Cereal snack bars have been promoted over the last decade as a healthy snack. In 1981, the cereal bar business in the UK was worth a mere £0.5 million. However, by 1988, cereal bar sales in the UK alone had reached the dizzy heights of £48 million.

Chewy cereal bars have recently become more popular than hard, crunchy cereal bars. The chewy bar often contains chocolate chips or has a chocolate or carob coating. Many cereal bar manufacturers claim that their bars are '100 per cent natural', 'full of goodness' or 'high fibre', which undoubtedly promotes a healthy image. In 1989, *The Food Magazine* conducted a survey on cereal bars and produced some surprising results:

Fibre None of the cereal bars in the survey met the government's Food Advisory Committee's 'high fibre' criteria, despite the fact that many of them were labelled 'high fibre'.

Calories Weight for weight, the number of calories in a fruit and nut bar can be similar to that in a chocolate bar so you should not be lulled into a false sense of 'not fattening' security! However, there are far more other nutrients in a fruit and nut bar than in a chocolate bar!

Fat On average, there is as much fat in a cereal bar as a popular chocolate bar. For example, the survey found that the fat content of cereal bars ranged from 27 per cent to 48 per cent, compared with a Mars bar, of which 35 per cent of the calories come from fat.

In the United States, many cereal bars contain palm or coconut oil, two highly saturated fats. Only a few UK manufacturers claimed they used healthier vegetable oils.

Sugar While the survey showed that most cereal bars contain large amounts of sugar, the majority of manufacturers avoid declaring the sugar content of their bars on the labels. Only four of the bars in the survey did declare 'added sugar', ranging from 20.4 per cent to 27.5 per cent. However, as the survey pointed out, this is only the 'added sugars' and does not include the 'hidden sugars' contained in the ingredients, for example in dried fruits or fruit juice, or the 'disguised sugars' like glucose syrup, fructose syrup, corn syrup, lactose, dextrose, molasses or honey.

The 'no added sugar' claim on labels implies that bars are low in sugar. However, *The Food Magazine* survey discovered that one fruit and carob bar bearing the 'no added sugar' claim actually contained 60 per cent sugar, mostly in dried fruit. Others in the range had an average 40 per cent sugar content. One of the carob bars contained more added sugar than the average chocolate bar.

The conclusion of the survey was that most cereal bars do not deserve their healthy image as they are too high in sugar and fat, and surprisingly low in fibre. This is another area where manufacturers appear to be misleading the public; once again, there is room for compulsory nutritional labelling in order to avoid further deception.

Old habits die hard

Foods and drinks containing sugar have a taste and texture which makes them highly desirable to many people. The taste for sweetness often begins soon after birth, and once children have developed a liking for sweet foods and drinks, it is difficult to wean them off. As children are particularly prone to tooth decay, it is advisable to try to establish good eating and drinking habits when they are very young.

Artificially-fed babies invariably have sucrose added to their milk, and their first solid food is often made more tempting by the addition of sugar. Little children are sometimes given sweet drinks in bottles, a near disaster as far as tooth decay is concerned. It is not widely appreciated just how much sugar there is in soft drinks. For example, how many teaspoons of sugar do you think there are in a can of Coca-Cola? You may be shocked to learn that in every can of Coca-Cola there are approximately 8 teaspoons of sugar (not to mention the caffeine). Chocolate biscuits, cookies and wafers are offered to children as rewards early on, so that they learn to associate good behaviour with a sweet reward. The table overleaf gives you the sugar content of some common foods.

'Hidden' sugar content of some common foods

Food	Weight/ Volume	Quantity	Amount of sugar grams	teasp
Boiled sweets	100 g	1 packet	95	24
Milk chocolate	50 g	1 small bar	26	6½
Plain chocolate	40 g	1 small bar	26	6½
Mars bar	52 g	1 bar	36	9
Bounty bar	52 g	1 bar	31	8
Fruit gums	20 g	1 small packet	12.5	3
Liquorice Allsorts	113 g	1 box	87	22
Lucozade	200 ml	1 tumbler	18	4½
Orange squash	40 ml	1 glass (diluted)	11.5	3
Rose hip syrup	40 ml	1 glass (diluted)	25	6
Ribena	40 ml	1 glass (diluted)	24	6
Coca-Cola	300 ml	1 can	32	8
Tomato ketchup	10 g	2 teaspoons	2	½
Chocolate biscuits	15 g	2 biscuits	7	1½
Chocolate digestive	15 g	1 biscuit	4	1
Rice pudding, canned	220 g	½ × 400 g can	16	4
Fruit pie filling	200 g	½ × 400 g can	34	8½
Fruit-flavoured yoghurt	150 g	1 small carton	14	3½
Fruit yoghurt	150 g	1 small carton	18	4½
Jam	15 g	2 teaspoons	10.5	2½
Honey	15 g	2 teaspoons	11.5	3
Ice cream	50 g	1 scoop	8	2
Jelly cubes	38 g	½ packet	24	6
Baked beans	225 g	1 medium can	10	6
Sweetcorn, canned	100 g	⅓ can	7.2	2

Children's sugary drinks

If you thought you were safe buying young children's fruit drinks labelled 'no added sugar', I am sorry to tell you that, once again, you have been misled. The 'added sugar' only refers to sucrose; children's drinks often contain naturally occurring glucose and fructose. I came across an interesting chart to demonstrate the extent of this deception. The chart is the result of a study by Professor Curzon and Drs Alemin and Duggal at the Department of Child Health at the University of Leeds School of Dentistry. They analysed a number of babies' and children's drinks and discovered that despite the fact that added sucrose levels were often low, as stated, the total sugar content was equivalent to as much as 4.1 teaspoons of sugar in the form of sucrose. These drinks, therefore, could cause just as much tooth decay as drinks actually containing sucrose. In fact, in 1991, Smith Kline Beecham paid an undisclosed sum in out-of-court settlements to 12 children whose parents claimed sugar caused damage to teeth.

Concentration of sugars in various baby and infant fruit drinks

Drink	Make	Dilution	Concentration (g/125ml)			Teaspoon equivalent of sucrose
			Sucrose	Glucose	Fructose	
Apple/cherry	R	*	0.875	4.82	8.69	4.1
Apple/plum/orange	R	*	2.00	5.62	6.75	4.1
'Moonshine'	L	*	2.75	4.86	6.62	4.1
Pear/peach	CG	*	2.81	2.13	8.63	3.9
Apple/blackcurrant	CG	*	2.37	3.25	6.63	3.5
Apple/orange	R	*	2.13	3.25	6.01	3.3
Apple/cherry	D	1:5	0.650	3.24	4.47	2.4
Apple/blackcurrant	D	1:5	0.0	3.80	4.50	2.4
Apple/orange	D	1:5	0.275	3.65	4.10	2.3
Summer fruits	CG	*	0.245	1.13	6.25	2.2
Baby 'Ribena'/orange	B	1:9	1.75	4.66	0.752	2.0
Baby 'Ribena'	B	1:10	0.075	5.07	0.725	1.9
Apple/blackcurrant	R	*	0.253	0.253	4.06	1.5
Gripewater	W	*	1.74	1.74	0.935	0.8

* Ready to drink
D=Delrosa B=Beechams R=Robinsons CG=Cow & Gate
W=Woodwards L=Libbys
Source: *The Lancet*, March 5, 1988 (page 539)

In fact, it seems that soft drinks provide as much as 14-21 per cent of added sugars in the diet of 11–12-year-olds. This was demonstrated in a dental study conducted in

Britain by Rugg Gunn and his colleagues in 1986. The study showed that cola-based drinks and blackcurrant drinks seem to cause the most damage. However, dentally speaking, other studies have shown that there is no major advantage in drinking natural fruit juices.

The solution to all this is to give only water or milk to children during the day, and water only at bedtime, in order to lower the risk of dental decay. Sweet drinks should be restricted to meal-times, after which the teeth should be cleaned. I appreciate that it's not as easy as it sounds – you have both advertising and peer pressures to contend with, not to mention the sweet shop on the way home from school.

It is unrealistic to think that you can deny children sweet food altogether, especially if they are already on the way to becoming sweet food junkies. However, you should bear in mind that many studies, including the famous Swedish Vipeholm study, have shown that eating sugary foods between meals leads to a higher tooth decay rate than when sugar is restricted to meal-times. It should be said that dental decay is an entirely preventable condition; it is an unnecessary burden we laden ourselves and our children with, at great cost to both our teeth and our pockets.

Operation Clean-up

If you want to wean children off junk food gradually, you can follow this plan over a period of weeks:

● Make sure they are cleaning all the plaque off their teeth at least twice a day. Stain the plaque for them with special disclosing tablets that you can buy in the chemist, so they can see where it is. (Remember to put some Vaseline on their lips first, though, to prevent them looking like one of Count Dracula's relatives afterwards!)

● If you happen to have a children's microscope at home, make a slide so that they can examine the plaque for themselves. The view down the microscope is pretty gross; it is almost guaranteed to halt them in their tracks. If no microscope is available, get a leaflet from your dental surgery with some pictures of the bacteria in plaque, and explain to them how plaque forms, and how tooth decay is caused by eating refined sweet foods.

- Establish a reward system for the children so that they get their 'sweet days' at the weekend, if they have earned them. This cuts down on the amount of sweet food they expect.

- Explain carefully about the effects that excessive consumption of refined sweet foods can have on their behaviour, their mood, and indeed their teeth.

- Get them involved with experiments to make healthy sweet-tasting snacks and puddings. Let all the family have a chance to give each dish a score out of ten.

- If you have a baby, please make sure you do not give him or her sugary drinks in bottles, and that dummies are not dipped in sugar or sugary drinks.

- Check what is on offer at school in the way of sweet foods. If possible, have a word with the head teacher or the parents' committee if you are concerned.

What to Eat Instead

There are a number of sweet-tasting foods that have virtually no relationship with tooth decay. Milk without added sugar is an example, as are fresh fruits. Foods like cooked rice, potatoes and bread are all low-risk in relation to dental caries.

There are plenty of foods which contain intrinsic sugars that can be substituted for the foods containing refined sugar. In Part II of this book, you will find lots of ideas for dishes to prepare and alternative snacks to try.

10

Artificial Sweeteners

FOR OVER 100 YEARS we have made use of a variety of artificial sweeteners. Their obvious attraction is that they provide the sweetness of sugar without any significant amount of calories. There are two types of artificial sweetener, the intense non-nutritive sweeteners and 'bulk' sweeteners. They do not, however, fulfil all the functions of sucrose. They do not provide the bulk needed to make ice cream, meringues and cakes; they cannot preserve foods in the way sugar does in jams and marmalades; and they cannot be caramelised for use in cooking.

Non-Nutritive Sweeteners

The use of intense non-nutritive sweeteners is mainly restricted to the sweetening of soft drinks, yoghurts and chewing gum, where the above characteristics of sucrose are not required. There are three main types:

Saccharin (Hermesetas, Natrena, Saxin, Boots Shapers, Sweetex)

Aspartame (Canderel, NutraSweet, Flix, Boots Shapers, Asda Table-Top, Tesco Table-Top, Superdrug Supertrim and Whitworth's Trimspoon)

Acesulfame Potassium (Hermesetas Gold, Sweetex Plus)

Saccharin

This sweetener was discovered in 1879 and is widely used. It is considered to be a safe and useful sweetener with a sweetness of between 200 and 400 times that of the equivalent weight of sugar. It is a cheap, calorie-free sweetener which is frequently added to soft drinks and squashes. It is also suitable for diabetics. It is recommended that no more than 25 mg per 1 kg (2 lb) of body weight is consumed per day. For a 55 kg (8½ stone) woman this would have the sweetness equivalent of 40 g (1½ oz) of sucrose. Unfortunately, saccharin has a significant after-taste which limits its use.

Aspartame

This sweetener was discovered in 1965 and its main advantage is its relative lack of after-taste. It is made up of two proteins, aspartic acid and phenylalanine, both of which are also found in much greater amounts in everyday foods such as milk, meat, cheese, fish, fruit and vegetables. It is metabolized in exactly the same way as other protein foods so it does not bring anything new to the diet.

Aspartame is about 200 times sweeter than sugar. The taste properties of aspartame are the closest of any sweetener to those of sugar. It is more expensive than saccharin and has the disadvantage that very occasionally adverse reactions occur, producing skin rashes or inflammation of the pancreas and abdominal pain, although some doubt has recently been cast on these side-effects. It is also thought by some researchers that there may possibly be a link between aspartame and neural tube problems in newborn babies. Children with the rare metabolic disorder, phenylketonuria (PKU) are unable to metabolize phenylalanine properly, and therefore have to restrict their intake of all foods containing it, including those containing aspartame.

The EC Foodstuffs Hygiene Committee has fixed the acceptable daily intake of aspartame at 40 mg per 1 kg (2 lb) of body weight. In theory, therefore, an adult weighing 68 kg (10 st 8 lb) could consume more than 75 sachets of Canderel or 16 cans of drink sweetened with aspartame daily.

Acesulfame Potassium (Acesulfame-K)

This sweetener was discovered in 1967. Chemically, it is very similar to saccharin and it tends to have a similar bitter after-taste. Its taste is 150 times as sweet as sugar. Studies in the United States have linked this sweetener to an increased cancer risk in some tests on rats, and to raised cholesterol levels in diabetic animals. The maximum recommended daily consumption of acesulfame potassium is 9 mg per 1 kg (2 lb) of body weight per day. For a 55 kg (8 ½ stone) woman this would provide sweetness equivalent to 74 g or 2¾ oz of sugar per day.

Nutritive Sweeteners

These are sugar substitutes that do provide some calories, though not as many as sugar. They have 'bulk' properties and are therefore useful in the confectionery industry, and they have the advantage of being less cariogenic, that is they cause fewer dental caries than sugar (sucrose) itself. Nutritive sweeteners are often combined with intense non-nutritive sweeteners. The 'bulk' sweeteners include Sorbitol, Mannitol, Xylitol, and altered glucose syrups (Malbit and others).

These 'bulk' sweeteners are known technically as 'sugar alcohols' or 'polyols', and they are close relatives of natural sugars. They retain their sweet, sugar-like, taste but most of them are not quite as sweet as sugar. Xylitol, however, is as sweet as sugar, and produces a cooling sensation when eaten. This is particularly noticeable if it is used in chewing-gum or toothpaste, and can enhance the pleasurable taste of mint. Interestingly, xylitol may actually help protect against dental decay, particularly if combined with fluoride.

Such polyols are also used in sweets and diabetic products. They do provide calories, but less calories than sugar – 2.4 kcals per gram, compared to sucrose which has 4 kcals per gram. Consuming moderate or large amounts of these polyols can produce symptoms of abdominal bloating, discomfort and diarrhoea, as they may not easily be absorbed and thus act as mild laxatives. Some adults and children are particularly sensitive. Those individuals who are intolerant of fructose (fruit sugar) should avoid sorbitol, which is converted to fructose in the body.

Mixtures of nutritive sweeteners and non-nutritive sweeteners are available, their purpose being to provide low-calorie sweetening agents with 'bulk' properties that do not cause so much dental caries.

The Disadvantages of Sweeteners

Sweeteners that are reduced in calories and cause less dental caries are to some degree a real advantage. However, they are not particularly good foods in themselves in the way that a stick of carrot or a piece of fruit is. They do not provide vitamins, minerals or fibre. For most people who would be healthier if they reduced their sugar intake, using sugar substitutes only serves to maintain a sweet tooth.

Alternative sweeteners are of most use to:

- people on calorie-restricted diets
- some diabetics
- children using sugar-free chewing-gums
- anyone who is incurably addicted to a sweet taste!

For the rest of us, it is better to concentrate on eating naturally sweet-tasting foods, rather than to make substantial use of added sugar or sweeteners.

Relative sweetness of sugars and sweeteners

Sucrose (table sugar) = 1.0

Saccharin	300	Lactose	0.3
Aspartame	200	Glucose	0.5
Acesulfame potassium	150	Fructose	1.5
Thaumatin	3,000	Mannitol	0.7
Cyclamate	30	Xylitol	1.0
		Maltitol	0.7

11
Sugar – The Reward?

Isolating the 'Final Straw'

THERE ARE a vast number of factors that push us to the nutritional edge. The complexities of life affect us all differently so it is crucial for you to isolate the reasons for the inconsistencies in your own diet. In my work at the Women's Nutritional Advisory Service (WNAS), I have discovered that people are sent off the nutritional rails by a range of circumstances. These are listed below; see if one (or more) of them applies to you.

- You dislike your shape so much that you feel there is no point bothering about altering it anyway
- You are upset because others have criticized your shape
- You have unrealistic expectations about your shape
- You feel isolated and lonely
- A special relationship has ended recently
- Your relationship is not running smoothly
- You have a poor self-image
- You sometimes experience waves of depression and apathy
- You experience Pre-Menstrual Syndrome and crave sweet food far more before your period is due
- Your life is filled with too much stress and pressure from work
- You are bored
- You are out of work and at a loose end
- You have financial worries

- There are family problems which you find difficult to deal with
- You are frustrated and lack direction in life
- You don't feel like you are achieving your goals
- You generally feel overwhelmed with life and all the problems that it presents
- You rarely eat your main meal in the presence of others
- You consume chocolate and other sweet food in secret
- You feel too busy or too tired to cook a proper meal for yourself once a day

Some Solutions

If one or more of the above factors applies to you, we have some work to do. If several apply to you, then you are going to have to work very hard at making adjustments to your life and perhaps even to your attitude. Here are some suggestions for you to chew on. They aren't designed to replace professional counselling, so if you feel somewhat out of your depth, it is advisable to seek professional help. We have found that these suggestions helped many women on the WNAS programme.

Your shape

To a large degree, your basic shape is inherited. It is normal for all women, at some time, to wish they were shaped like Julia Roberts or Sophia Loren, and for men to wish they were a cross between Arnold Schwartzaneger and Richard Geer. However, we have to make the best of what we have got.

If you really are overweight, you should work hard once and for all to shed the excess pounds. Do not seek out a fad diet or a very low-calorie diet that will only serve to slow down your metabolism in the long term, and consequently make it even harder for you to lose weight. You must find a diet that suits you, and one on which you will feel well without going hungry. If you haven't found a diet to suit you, you might like to try the Vitality Plan, the details of which are in a book of the same name that I co-wrote with my husband (see page 183). The main aim of the book is to help you find the right kind of diet for your body, so that you not only lose weight, but feel well and 'vital'. It provides valuable education about diet, and will help you to isolate the diet to follow for the rest of your life. Once you have reached your target weight, you can increase your calorie intake and continue with the diet. Of course, you can deviate from it when you go on holiday or at Christmas, for example, but at least you will know how to put it all back together again when the party is over.

Exercise will help to speed up your metabolism so that you lose weight more easily. You need to exercise for at least 40 minutes three or four times a week, or initially to the point of breathlessness. Exercise will also help to improve your shape. If you choose the kind of exercise you like, there is a greater chance that you will stick to the exercise programme.

The style of clothes you wear can also have a bearing on how fat or thin you look. Choose your styles carefully to make the most of your figure.

If someone has unkindly criticised your shape, or even made comments which they genuinely believed were for your own good, do not take them too much to heart. When you have shed a few pounds and are feeling in the peak of health, you will have greater self-esteem and such comments will not bother you.

Facing the world alone

For many of us, being alone is never easy, no matter what stage of life we are at. It is all too easy to lock yourself away in a TV-dinner-existence, but in the long term this could be a recipe for disaster. Withdrawing from the world is certainly not the answer, even though it might be the preferred solution at the moment. It may be that you don't find it easy to develop relationships, or that you have recently separated from your partner. Whatever the situation, you must look at ways of making friends, and getting out of the house.

If you hate being in a sink-or-swim situation socially, don't force yourself to go to bars or clubs. Get the prospectus for your local authority evening classes. You may be pleasantly surprised at just how many options there are to choose from. Have a go at rekindling an old hobby, or take a course on a subject that you would like to learn more about. In addition to this, there are appreciation societies, walking societies, health and fitness clubs and a variety of voluntary organisations that would be only too grateful for your help. Hopefully, before too long, your social schedule will be much fuller, and you won't have the time to sit in alone.

Many of us suffer to some degree from a poor self-image. The fact is that you have to give people the chance to get to know you before they can appreciate your qualities. Most of us have something special to offer or contribute once we put our minds to it. You may have to talk to yourself on a regular basis to get you through difficult times, but the chances are you will eventually come to realise that you are as able and accomplished and nice to know as the next person.

Love on the rocks

A stormy relationship, or one that is failing miserably, can be extremely draining and upsetting. If you have been with your partner for some time, and feel that the relationship is salvagable, then I suggest you work at putting it back together again. Lots of honest communication goes a long way. If you can't face it on your own, then ask a friend or relative to help you both talk through your difficulties. If that fails then it might be an idea to seek expert advice from an organisation such as Relate (see page 187).

When a relationship ends, even if it is by mutual consent, a void is often created. Apart from the natural distress and loneliness, a confidence crisis sometimes arises. Make sure you keep in company as often as possible. Do not sit alone moping, as this may well be the time that you take to comfort-eating or even bingeing.

Make time to plan how you will fill the social time you now have at your disposal. Immerse yourself in an old hobby or some new activity you have been wanting to try for some time. Surround yourself with your friends and family, and above all, think about the next chapter of your life which is about to begin. Try to regard it more as a new adventure, rather than a daunting prospect.

The monthly blues

If you suffer from Pre-Menstrual Syndrome (PMS), you may well find that, to some degree, your cravings for sweet food become worse pre-menstrually. Having helped tens of thousands of PMS sufferers over the years, it is only too real to me how severe the problem can be for up to two weeks each month. The Sugar Craving Syndrome can be accompanied by fatigue, dizziness, palpitations, lack of concentration and even depression.

There is a specific diet that you can follow to help alleviate the symptoms of PMS. Further details are in a book called *Beat PMT Through Diet* (see page 183). If you feel that you need extra help, you should either consult your doctor or a nutritional adviser. The WNAS runs clinics in the UK, but also has a postal service available to women who cannot get to them. The address is on page 185.

If you suffer with depression but do not know why, then it is possible that your symptoms are diet-related. Have a go at the Vitality Plan (see page 183), to find out if there are any foods or drinks that cause your feelings of depression. Plus, it might help you to exercise three or four times a week and to keep yourself so busy that you haven't the time to be depressed. You could offer your services as a voluntary helper to some less fortunate group in your community, or perhaps even do some fund-raising for a worthwhile charity. If you don't feel you are making progress, then you should consider getting some expert counselling.

Stressed out and overwhelmed

There is nothing quite like chronic stress to knock you for six. We all have different tolerance levels under pressure; I usually find that some degree of stress is healthy as it stretches me. However, there is a fine line between stress and *di*stress. When you have reached the realms of regular distress, whether it be at home or at work, you must try to stand back from the situation and to make some adjustment in your life.

If your job is too pressured for your liking, then perhaps it is time to take stock and examine whether you think it is worth it in the long run, or whether you should be looking for a more acceptable position. You can only continue under excess pressure for so long before something, be it physical or mental, snaps. You would be wise not to let the situation continue that long.

Drinking excess coffee, smoking too much or pigging out on chocolate and sticky buns are all phenomena of long-term distress. In order to reverse this established habit, you may need to take a few days off work or perhaps go off somewhere for a week, where your schedule is relaxed and different. We usually find it easier to adjust our habits when we are out of our usual environment. Once you have broken the pattern to some degree, you should be more able to contront the situation and to make the necessary adjustments once back in your usual routine.

If you feel your life is becoming too stressful, it might be useful to give some thought to making new policies, and new routines, to avoid being stretched to the point of distress. Make sure you take a break in the middle of the day; get some fresh air and some exercise. You might even consider taking up yoga or having a regular massage to avoid getting too stressed out in the future.

Time on your hands

Being at a loose end is definitely the other end of the spectrum, and while many busy people may be envious of you, it can be a very soul-destroying time. It is very common to eat out of boredom, or because you feel that life has nothing to offer. It is vital to fill your time with things that you feel are worthwhile. If you are not working, don't sit at home waiting for something magical to happen; go out and do something rewarding. If you are unemployed temporarily, there are many voluntary organisations that would be glad of some help, or you could do some decorating or gardening for an old person. Relieve a young mum of her kids for a few hours each day, or go on a course that will provide you with new skills. There are so many options, but you usually have to seek them out rather than wait for them to come to find you.

Lacking direction in life

Wandering aimlessly through life may be fun for a while when we are young, but it may not do much for your morale in the long term. Most of us have, at one time or another in our lives, consciously or unconsciously had a goal that we were working to achieve. If you are feeling frustrated, and wonder what life is all about, it may be time for you to sit down and work out a direction that you can follow in order to pursue a goal that you consider to be worthwhile. The satisfaction you get from actively working to achieve your goal will improve your self-esteem and give you much more purpose in life. It may be that when you have worked out your goal you will need to change direction in life; don't be afraid to make changes that make you feel more satisfied with life and with yourself.

Making ends meet

They say that 'money makes the world go round' and sadly, in many ways, this is true, especially these days with the economic climate making things tight all round. It can be very depressing and worrying to find money is in short supply, especially if you know that your income is less than your outgoings. Psychologically, we often turn to sweet foods for comfort, but in fact they are little compensation in the long term if you look at it logically.

Planning your finances ahead is essential. If you are truly spending more than you make, then you have to make some serious adjustments to your lifestyle on an emergency basis. There are various ways to go about redressing the balance, depending on your circumstances. If you are single, then you need to choose from the following list:

- A pay rise at work
- A new job paying a higher wage
- Some part-time work to supplement your income
- Examine what you spend your money on and try to cut corners
- Move to cheaper accommodation
- Rent out a room in your house
- Perhaps consider taking on a slightly higher mortgage, which could possibly release a small amount of capital to settle any debts that have accrued

If you are married or co-habiting, then there may be more leeway. The chances are that you will be able to think of ways to earn more and also to cut down on your existing outgoings together.

Once you have made a new financial plan and you have put it into action, you will need to keep accurate records and do a thorough check each month to see whether the books do in fact balance. It is an interesting exercise to make a graph of your income and outgoings so that you can visually see that there is a healthy difference or otherwise. Most of all, don't sit and stew alone. If you feel overwhelmed by your financial problems, ask a friend or relative to give you some help, and if all else fails, seek advice from an expert.

The ups and downs of family life

It is almost inevitable that there will be times in family life when things are not running as smoothly as you would perhaps like. There may be worries over children who are not behaving in the way you feel they should, or not keeping the company or hours you approve of. When communication breaks down, it can be both upsetting and frustrating. We often resort to self-recrimination at times like this, when in fact the real reason for the problem is nobody's fault. Comfort-eating at this time may help in the short term, but in the long term there is no substitute for improved communication between yourself and the children. Talk out the difficulties, even if they take weeks to resolve. There is usually a compromise that can be reached; if not, expert advice of the relevant sort should be sought.

We all go through the traumas of worrying over sick relatives, and ultimately coming to terms with the passing of aged parents. It is never easy and at times of immense emotional stress we often turn to sweet food or alcohol for comfort, just to get us through the very difficult times. In the long term, however, you would be doing yourself a great injustice if you continued with these emergency habits. You must take care of yourself at all costs. Actively make a plan to exercise daily, eat healthily, have a change of scene for a while, and to talk out your problems or your grief with a partner or close friend. Whatever you do, make sure that you don't bottle up your feelings; it will take you so much longer to solve the problem or come to terms with the loss if you do.

The lone eater

If you actually prefer eating alone, the chances are that you are already on the slippery slope towards an eating disorder. Many people who suffer with anorexia nervosa (the wasting disease) or bulimia (bingeing) prefer to eat alone. In fact, they often refuse to eat in company. (See page 82 for more information about these disorders.) Whatever you do, make sure that you eat with someone. If you live with your family, make a rule that you all eat together for as many meals as possible. If you live alone, it is obviously more difficult. However, you can make reciprocal arrangements with friends to eat together.

The same rules apply to eating sweet foods – only eat them in company. By all means, keep a supply of goodies in the cupboard to avoid feeling that there is nothing available to satisfy your cravings, but don't consume large amounts of sweet foods when you are on your own. The chances are that this rule will make you think twice before you help yourself to the second, third or subsequent portion of pudding, plate of jam tarts, or chocolate bar! Give it a try. You will see what I mean.

It is important to stand back and to try to take stock of the situation from an external point of view. As our eating habits are so closely related to our social circumstances, it makes sense to have a personal spring-clean across the board. If you can spot troubled areas and make adjustments yourself, so much the better. If not, don't be shy to seek professional advice; we are all of us fallible at times.

12

Mental Health in the Balance – Eating Disorders

THE WAY WE FEEL mentally may have a direct effect on our shape. It is important to clear out your 'psychological cupboard', as it were, before proceeding with the new dietary regime recommended in this book. It stands to reason that if there are psychological or emotional barriers or reasons for bingeing, you may well not succeed with any diet you attempt, now or in the future.

Without wishing to worry you unduly, it must be appreciated that there is a very fine line between being a 'chocoholic', or a part-time binger, and having a real eating disorder. The eating disorders, anorexia nervosa and bulimia nervosa, are both serious conditions which can have extremely unpleasant consequences. They represent a significant health risk in certain groups of the population. Included in these high risk groups are models, members of the acting profession, dancers, and others who are overly concerned about their body weight and shape.

Anorexia is a condition whereby the drive to be thin is all-consuming. There is usually a dissatisfaction with the body shape and a distortion of how it is perceived by others. This condition is often manifested by strict dieting, obsessive exercising, and vomiting or laxative abuse after eating. Bulimia sufferers share the concern about their body weight and shape, and the drive to be thinner, coupled with a phobic fear of fatness. The most obvious symptom of bulimia is bingeing, in other words the consumption of large amounts of usually not very nutritious foods in a short period of time with a sense of loss of control. Episodes of bingeing are often followed by strict dieting or by enforced vomiting or laxative abuse. Bulimia sufferers are not necessarily overweight, although they may well be, and they often indulge in excessive or vigorous exercise when they are not bingeing.

It is reckoned that approximately 1 per cent of women in the 15-40 age group suffer from anorexia, while 2–3 per cent of the same age group suffer from bulimia. These disorders do also occur in men, but much less frequently, probably because, in general, men tend to be less concerned about having a slim body image. A further 5 per cent of women suffer from a sub-clinical case of either syndrome, which means that they have some of the symptoms of the condition, but to a lesser degree.

While it is thought that these illnesses most typically occur in the 18–25 age group, they can also occur in the early teens or even well into adulthood. They are becoming an increasing problem amongst younger girls and should very definitely be watched for and nipped in the bud. These conditions were previously considered to be disorders of the upper and middle classes. Over the last 15 years, however, the differences in social class distribution seem to have disappeared.

Both of these conditions are serious and should not be taken lightly. Follow-up studies suggest that there is a death rate of about 15 per cent over a period of 30 years. If, after reading about anorexia and bulimia, you feel that your condition may be more severe than you had previously thought, it would be best if you went to see your doctor for a check-up and perhaps consulted one of the special help organisations listed on page 187.

The Male Approach

As already stated, men do not seem to suffer from the same drive to be thin that can lead women into these serious eating disorders. The reasons for this are complex;

basically women are under great pressure in today's world to conform to a particular image. Men are not subjected to the same pressure.

Whereas women tend to binge on sweet food in secret, men often seem overtly proud of the amount of chocolate and similar foods that they are able to demolish. Here are two typical cases that were related to me with pride!

Bobby Mills is a 26-year-old man who lives in Sussex. He has been consuming massive amounts of chocolate since he was six years old.

'I can't stand fruit and vegetables or meat or fish and I rarely eat any salad. I love chocolate and eat it as often as I can. I start my day with two Mars bars on waking.

I get through an unlimited number of chocolate bars per week, depending on how many I can get my hands on. I also eat whole family-sized Swiss rolls and gâteaux all to myself whenever I get the opportunity, and I eat loads of chocolate biscuits and choc-ices and drink litres of cola drink.

I remember the time I had a bet with some friends that I couldn't eat seven Mars bars at one sitting. I did actually manage to eat all seven Mars bars with ease in 20 minutes. Later, when I got home, I consumed another two as I was feeling peckish! They never did pay me for winning the bet.

Maryland cookies with chocolate chips are another of my weaknesses. I often eat a whole packet in 10 minutes without any problem and could certainly eat another packet without forcing myself. I get through at least 3 litres (5¼ pints) of Coke per week and three large cartons of orange juice.'

Bobby hardly eats any nutritious food at all. He lives on a diet of chocolate, cakes and fizz. Although he has never before stopped to consider how much his sweet tooth is costing him financially per month, he agreed to do so. He calculated that in an average month he spends £52 on sweet foods and drinks. In the long term, health-wise, it will cost him a great deal more!

PART II
BEATING THE CRAVINGS

13
The Good News

THERE *IS* a specific way to go about overcoming the cravings you have for sweet food in a controlled manner. However, before embarking on this new programme you need to isolate precisely where you have been going wrong in the past.

Recipe for Disaster

Read through the following scenario and consider whether you have been guilty of any of its aspects:

> The day starts with either no breakfast or just a couple of cups of tea or coffee and a few cigarettes. This may stimulate your body's metabolism and increase your energy levels for a short while, but it is likely that you will run into problems by mid-morning. By then your energy levels will begin to fall, and it is likely that symptoms of nervousness, anxiety, palpitations, light headedness, headaches, hunger and cravings for food, particularly sweet foods, will set in. Another round of tea or coffee with sugar, or with a sugary or chocolate snack, may delay the symptoms for a while, but not for very long, as these habits are a recipe for disaster.

The Way Out

The first step towards overcoming your cravings for sweet food is to eat wholesome food regularly. This does not mean that you have to do without sugar, as long as the sugar is only there to enhance nutritious food. Ideally, you should be eating 'little and often' in order to keep your blood sugar levels constant. Keeping your blood sugar level on an even keel is a major part of the key to success. The diet includes not only breakfast, but also wholesome snacks between meals, at least to begin with. This means that you could be eating as many as six times a day to start with. You will need to pay extra attention to your oral hygiene while eating so regularly by increasing the number of times you clean your teeth thoroughly each day to at least three.

You are probably thinking 'Eat six times a day? I'll get as big as a house.' This is not actually so. The diet is based on an intake of approximately 1,500 calories per day. What you will find is that the portion sizes of the main meals are smaller than they would otherwise be, to compensate for the snacks. Men should automatically increase their portion sizes of the protein and vegetables by one-third in order to consume sufficient calories.

The diet is arranged in daily steps, with a specific menu and recipes for each day. If you are vegetarian, or you are not keen on the suggestions for the day, you can substitute one day for another. Accompanying each menu is a 'Fact for the Day', which will hopefully expand your knowledge about sweet foods in relation to your body.

Before beginning the diet, you will need to have read the section entitled *Sugar – The Reward?* in great detail. It is important to identify and, ideally, eliminate any aspects of psychological dependence you may have on sweet foods. If necessary, go back and read the section again.

Work It Out

The third thing I am going to ask you to do is to work out an exercise programme for yourself, which should consist of three or four sessions of exercise a week lasting for a minimum of 30 minutes each. If you haven't been exercising regularly, you may find that you need to push yourself initially to get started. Once you begin to feel the benefits of exercise, that will be all the motivation you need to keep you going.

The purpose of exercise is to keep you fit and to speed up your metabolism. A considerable amount of research has been carried out on exercise in relation to mood swings and depression. The conclusions are positively in favour of regular exercise. It doesn't much matter what kind of exercise you undertake, as long as you adhere to these two rules: you must enjoy the exercise you choose, and you should exercise to the point of breathlessness each time (in other words, the exercise should be 'aerobic').

You should not treat your exercise as a competition. The only person you will be competing against will be yourself, so if you haven't exercised for a while, ease into it gently. Go for brisk walks to start with; you may find that you get puffed out after only 10–15 minutes. Do not overdo it; it is far better to build up gradually over a period of weeks. If you have any current health problems, you should check with your doctor before you embark on an exercise programme.

The types of aerobic exercise you can choose from are brisk walking, jogging, swimming, skipping, cycling, squash or a formal work-out in an exercise class. You can even purchase one of the exercise videos and follow the routine at home. This requires much more self-discipline than exercising with a partner or attending a class, but it is a good way to get started if your schedule doesn't permit you to get to a class on a regular basis.

A Nutritional Prop

The fourth and final part of the programme is that you should take specific nutritional supplements, at least to begin with, as a short-term nutritional prop. There are several important vitamins and minerals associated with blood sugar control. They can be taken individually or in the specialised combination supplement, Sugar Factor (see page 58). Sugar Factor contains all the important vitamins and minerals you need to help overcome your cravings for sweet food. It saves having to purchase and take several different tablets at once. Details about where this supplement is available are on page 184.

Evelyn Levene is a 41-year-old company director who is married with three children. She experienced cravings for sweet food for over 10 years:

'I used to crave chocolate, especially in the evenings and at night, plus every day I would eat at least a bun, a doughnut and several biscuits. In the evenings I usually got through at least two chocolate bars. All in all, I reckon I spent in the region of £20 per month on chocolate and junk food. I felt I needed the sugar as it gave me a "pick me up".

I suffered from loss of concentration, like I had cotton wool in my head. I was short-tempered and didn't feel on top mentally. After giving way to my cravings I felt like a "pig".

I had a four-month course of treatment at the Nutritional Advisory Service Clinic. I followed the diet suggested and took the Sugar Factor supplements. To my joy, my cravings for sweet food disappeared during the first month. I used to have to re-schedule appointments on bad days because I couldn't concentrate. Now I feel like a completely different person. I can hardly believe it.'

Clare White is a 31-year-old mother of two who works as a housewife.

'I couldn't stop thinking about sweet foods, and I used to eat all day. Needless to say, I had a weight problem. I could never stop eating sweet foods long enough to give myself a chance to lose weight. I had uncontrollable cravings. I would eat chocolate, biscuits, honey, jam, syrup – anything I could find in the cupboards. I used to search the cupboards in secret to find sweet food, and eat it quickly in the kitchen so that my family did not see.

When I had finished eating I felt bloated, lethargic and often nauseous. I felt that I was out of control, which made me feel very depressed. I knew I would put on weight and I felt guilty.

I contacted the Nutritional Advisory Service for help. It took me about two months on the diet to overcome my cravings for sweet food. I am a lot more in control now. I found the Sugar Factor tablets particularly helpful in the early stages. I think that breaking the pattern of eating is the most important thing. If I ever feel like slipping into old habits now I try to manage a whole day without anything sweet, except for fruit, which I find helps tremendously.

I am no longer obsessed with sweet food and I have a great deal more energy, which I really value as I have two small children.'

Before You Begin

Make sure you have time to go shopping before you attempt to start the diet, and go shopping with a full stomach *please*. The worst nutritional crimes are committed on an empty stomach! Clear all the forbidden foods out of the cupboard you will be using so that you are not constantly tempted to indulge. If you suffer even vaguely with pre-menstrual sugar cravings, do not begin the diet until after your period has started. It is probably better to begin the diet over a weekend, if you have weekends off, as this will give you a little time to get to grips with it. You may experience some withdrawal symptoms during the first week or two while your body is adjusting to the new eating habits. Your body may not like giving up sweet food, cola-based drinks and chocolate without a fight. Fore-warned is fore-armed: you may experience headaches, irritability, fatigue, insomnia or even depression initially. Stick with the diet despite these unwanted symptoms, if they occur. It will only be a matter of days before they pass and leave you feeling better than you have done for years.

14

Making a Workable Plan

SADLY, there is no magic pill or potion that you can take to overcome your cravings for sweet food but there is a plan you can follow on a daily basis in order to help normalise your blood sugar levels and combat the cravings. Before embarking on the plan you need to give a little thought to all that will be required of you, so that you feel mentally prepared and sufficiently committed to follow it through. After all, there is little point in even beginning if you do not fully intend to stick to it.

What the Plan Consists of

1 Following a diet high in the minerals chromium and magnesium, and in vitamins B_1, B_2, B_3, B_6 and C. The 30-day diet outlined has been specially tailored to provide these nutrients in concentrated amounts. This diet should be followed on a daily basis as closely as possible. (Additional nutrient food lists can be found on pages 98–104.)

2 Taking regular exercise (at least 3–4 sessions a week) in order to keep your metabolism ticking over at an optimum rate, and to keep you feeling fit and healthy. Choose the kind of exercise you like and establish routine times to do it.

3 Taking specific nutritional supplements with a high content of the minerals chromium and magnesium for the first 3–4 months.

Preparatory Steps

1 You will need to set aside a little time to prepare for beginning the diet, preferably a free evening or part of a weekend, if you are working.

2 On the following pages are lists of foods and drinks that you can consume freely, those that can be consumed occasionally and those that are forbidden. The first thing you should do is read through the lists and make a shopping list as you go along.

3 Next, read through the nutrient-rich food lists on pages 98–104, and make a note of all the foods you like.

4 Once you have a clear idea as to which foods and drinks will be on your approved list, you need to go through your kitchen cupboards, fridge and freezer with a fine-tooth comb, setting aside all the forbidden foods. It is preferable to give the forbidden foods away so that you avoid the possibility of temptation in the coming weeks. If the foods are expensive and you cannot bring yourself to part with them, then put them in a far corner of the freezer, or in a cardboard box on a high shelf so that they are out of everyday sight and mind.

5 Shopping is next on the agenda, preferably not with an empty stomach. Make your list before you leave home. Make sure you are not hungry before you set off, and promise yourself you will stick faithfully to the list and not be tempted to buy any forbidden items.

6 The diet will provide you with approximately 1,500 calories per day. You can adjust the portion sizes, depending on the size of your frame. You may well find that, if you are overweight and hoping to lose weight while on this diet, you will need to prepare slightly smaller portions than suggested. Weigh yourself before you begin and make a note of your starting weight. It is not advisable to weigh yourself daily as your weight will fluctuate from day to day; it is better to weigh

yourself weekly, in the same state of dress or undress, and try to use the same scales each time as different scales may give a slightly different result.

7 Finally, the last ingredient you need will be a little will-power. You have probably tried several times in the past to give up the sweet foods and drinks that you cannot resist. These attempts have obviously not met with much success or you would not be about to embark on a fresh attempt. The first week or two may not be easy for you. Once 'addicted' to certain foods and drinks, the body does not like to part with them without putting up a good fight. You may well experience withdrawal symptoms initially, which can last anything from one day to two weeks. The withdrawal symptoms may be in the form of headaches, tiredness or irritability. If you are lucky, you will not experience any withdrawal symptoms at all, or only mildly. However, if you do feel unwell for a few days, bear in mind that this is a temporary phase and that a sense of well-being will follow shortly. The worst thing you can do is to give up when you are halfway through the withdrawals, because the day will come when you have to go through the whole procedure again. If you look at it from that point of view, you may as well see it through now!

Dietary Goals

The goals of the diet are:

1 To reduce your consumption of added sucrose, glucose and fructose. This involves limiting your calorie intake from added sucrose, glucose and fructose to 5 per cent of total calorie intake. On a 1,500 calorie diet, this will be 75 calories, equivalent to 20 g (4 teaspoons) sucrose daily.

2 To establish a healthy eating pattern
   (a) by eating regularly, and
   (b) by eating a nutritious, well-balanced diet.

3 To reduce the swings in blood glucose level that may occur in the normal day.

4 To limit calorie intake to approximately 1,500–2,000 calories per day, depending on the size of your frame.

Versatility of the Diet

The diet section consists of menus for each of the 30 days, together with suggested recipes. If you enjoy cooking and have the time to prepare meals, you might like to follow the diet precisely as it is laid out. If, however, like many of us, you are working away from home and lead a hectic lifestyle, you can modify the diet. Just a glance at each day's menu will tell you how much cooking is necessary; those recipes marked with an asterisk are included in the book. You can replace hot meals with cold meat or fish and salad or raw vegetables. Alternatively, you can select the hot meals from the diet that you find easy to prepare and use them more regularly than suggested. As both the nutritional and the calorie content for each day fall within a similar range, it is easy to juggle the meals to suit your lifestyle.

Vegetarians

A number of vegetarian meals have been included in the diet, although it has not been constructed exclusively with vegetarians in mind. If you prefer to follow a vegetarian diet, you can substitute the meat and fish meals for vegetarian meals suggested elsewhere in the diet. It means you will have slightly fewer recipes to choose from than meat-eaters, but the diet will still be healthy and interesting.

Desserts

You may be pleasantly surprised to see that a sweet dish is suggested for each day. The idea is not to deny you sweet food, but to get your body used to moderation.

Again, if you are not keen on preparing desserts each day, you can substitute fresh fruit for the suggested dessert. Perhaps you could prepare some of the desserts at the weekend for a treat. Try not to miss them completely as many of them are quite delicious; you can have them either at lunch or dinner.

Snacks Between Meals

The diet includes mid-morning and mid-afternoon snacks. The purpose of this is to maintain your blood glucose at an optimum level. Those who have a slow metabolism, and are very prone to weight gain, may feel that the diet allows too much food each day! If this is the case, then you can drop one or both of the snacks, provided you do not feel hungry as a result. You will find some suggested snacks listed on page 107.

Portion Sizes

The calculations of nutrient and calorie content are based on average portion sizes. For example, each serving of meat is appoximately 100–150 g (4–5 oz), and each serving of fish is in the region of 150–175 g (5–6 oz). A medium-sized jacket potato should weigh roughly 100–150 g (4–5 oz), and the same goes for a portion of cooked rice (about 25g/1oz uncooked). The salad recipes indicate how many portions each one makes, so it should be easy to work out the approximate size each portion of salad should be.

Where possible, unless you are keen to maintain your weight or even gain weight, I suggest you use low-fat cheese and yoghurt, and low-calorie mayonnaise and salad dressing. Where recipes include milk, semi-skimmed milk should be used as opposed to full-fat milk.

Beverages

It is important to drink plenty while on the diet. As you will see from the list on page 96, bottled water and herb tea can be consumed as often as you like. It is best to drink fruit juice with a meal as the vitamin C content is high; this will increase your iron absorption, a particularly important factor for menstruating women.

It is best to avoid cola-based drinks, and other fizzy drinks, sweetened or unsweetened, while on the diet. If you fancy a fizzy drink, try mixing carbonated water with a little fruit juice yourself, or buy carbonated apple juice or spring water with a dash of added fruit juice.

Tea and coffee should be restricted as much as possible, as you should be avoiding the caffeine and tannin they contain. If you cannot do without the odd cup of tea or coffee, make sure you restrict your intake to no more than two cups of either per day.

How to Proceed After the Initial 30 Days

After following the programme for 30 days you should be feeling a little more healthy generally, and will perhaps be a little trimmer. More importantly, though, you will feel you have got to grips with your sugar cravings. After many years of working with people suffering from these cravings, we have come to the conclusion that there are no hard and fast rules about how long it takes to feel that the problem has been completely overcome. Whilst some people feel that three or four weeks is sufficient to break the craving pattern, others find that it takes three or four months. Only you will be able to judge how much progress you have made in 30 days.

If you haven't followed the diet as closely as you might have, you can always start again. On the other hand, you may well be satisfied with your efforts and pleased with your progress. However well you feel you have done, I would suggest that you continue to follow the dietary recommendations, and possibly the diet itself, for another few months, until you feel you have genuinely overcome your body's cravings for sweet food. It sometimes takes as long as four months to change the habits of a lifetime.

When you consider the long-term benefits of the diet, however, I am sure you will agree it is a very worthwhile investment of time and effort.

I hope you enjoy following the diet and that your new eating habits prolong your health and well-being in the future.

Dieting Tips

- Start when you are ready
- Plan ahead
- Go shopping first
- Keep to your shopping list
- Always shop after you have eaten
- Eat a variety of foods
- Eat fresh foods whenever possible
- Eat regularly – at least three meals every day
- Eat from a small plate
- Give yourself time, relax and enjoy your food
- Savour the flavour – eat slowly and chew well

Foods and drinks that can be consumed freely

Salad: Tomatoes, Cucumber, Celery, Lettuce, Watercress, Radishes, Chinese leaves, Spring onions, Beetroot

Green vegetables: Broccoli, Spinach, Spring greens, Kale, Brussels sprouts, Courgettes, Cauliflower, Cabbage (green, white and red), Green beans

Additional vegetables: Carrots, Sweetcorn, Onions, Peppers (red and green), Chillies, Beans, Pulses

Rice and Rice cakes

Herbs and spices

Bottled water

Herb tea

Foods and drinks that can be consumed in moderation

Lean meat

Fish

Poultry

Eggs

Low-fat cheese

Root vegetables: Potatoes, Parsnips, Turnips, Swede

Fresh fruit

Nuts and seeds

Bread and Ryvita

Semi-skimmed milk

Butter

Low-fat yoghurt

Small amount of single cream

Breakfast cereals

Canned fruit in natural juice (in small quantities)

Dried fruit in 25 g (1 oz) portions (once or twice a week only)

Low-calorie mayonnaise and salad dressing

Cold pressed oils, e.g. sunflower, safflower, sesame or walnut oil

Chewing-gum sweetened with xylitol

Beverages: Tea and coffee (no more than two weak cups a day)

Alcoholic drinks (no more than three units a week; a 'unit' of alcohol is one glass of wine, one measure of a spirit or half a pint of beer or lager)

Fruit juice (two glasses a day)

Foods and drinks that should be avoided whenever possible

Sugar and honey (unless otherwise stated in the diet)

Glucose

Cakes

Biscuits

Pastries

Chocolate

Sweets

Mints

Toffees

Chewing-gum (unless sweetened with xylitol)

Saturated fats (animal fats), other than small amounts of butter

Fried foods

Beverages: Cola-based carbonated drinks, cans of sweetened or unsweetened carbonated drinks, more than the stated alcohol, tea or coffee allowance

Nutritional Content of Food

The following lists detail the foods that contain good amounts of each of the vitamins and minerals that will help combat your sugar craving.

Read the lists through carefully to see which nutrients your favourite foods contain. You can then concentrate on foods with good quantities of the right nutrients in your diet.

Foods containing vitamin B

Cereals
Wholemeal flour
Wheat bran
Soya flour
Brown rice
Corn Flakes

Meat
Lamb's liver
Pig's liver
Bacon (lean)
Gammon (lean)
Beef (lean)
Minced beef
Lamb breast (lean)
Veal
Chicken
Duck
Turkey
Steak
Pork chop
Rabbit

Fish
Cod
Salmon
Plaice
Herring
Kipper
Mackerel

Pilchards
Tuna

Fruit
Bananas
Apricots (dried)
Prunes
Raisins

Vegetables and pulses
Spinach
Butter beans
Haricot beans
Mung beans
Red kidney beans
Chick peas
Peas
Broccoli florets
Brussels sprouts
Cabbage (red)
Cauliflower
Avocado
Leeks
Potatoes

Nuts
Hazelnuts
Peanuts
Walnuts

Other

Tomato purée	Beef extract
Bovril	Milk (skimmed)
Marmite	Milk (fresh whole)

Foods containing chromium (per 100g/4 oz)

	mg		*mg*
Meat		**Fruit**	
Calves' liver	55	Apple	14
Chicken	15	Banana	10
Lamb chops	12	Orange	5
Pork chops	10	Strawberries	3
Eggs		**Vegetables**	
Hens' eggs	16	Cabbage	4
		Carrots	9
Fish		Fresh chilli	30
Scallops	11	Green beans	4
Shrimps	7	Green peppers	19
		Lettuce	7
Cereals		Mushrooms	4
Rye bread	30	Parsnips	13
		Potatoes	24
Dairy		Spinach	10
Milk	1		
Butter	13	**Other**	
		Brewer's yeast	112

Foods containing magnesium (per 100g/4 oz)

	mg		*mg*
Cereals		**Dairy**	
Wheat bran	520	Dried skimmed milk	117
Wholemeal flour	140	Fresh whole milk	12
Oatmeal (raw)	110		
Porridge (rolled) oats	30	**Meat**	
Soya flour	290	Beef (lean cooked)	11
Wholemeal bread	230	Lamb (lean cooked)	12
Muesli	100	Chicken meat (roast)	24

	mg		mg
Fish		**Nuts**	
Cod (baked)	26	Almonds	260
Herring (grilled)	32	Brazil nuts	410
Kipper (baked)	48	Walnuts	130
Pilchards (canned)	39	Peanuts	180
Salmon (steamed)	29		
Sardines (canned in oil)	52	**Vegetables and Pulses**	
Winkles (boiled)	360	Butter beans (boiled)	33
Crab (boiled)	48	Haricot beans (boiled)	45
		Mung beans (raw)	170
Fruit (raw)		Chick peas (cooked)	67
Pineapple (fresh)	17	Spinach (boiled)	45
Apricots (fresh)	12	Potatoes	
Apricots (dried)	65	(baked with skins)	24
Bananas	42	Avocado	29
Blackberries	30		
Dates (dried)	59	**Beverages/drinks**	
Figs (dried)	92	Herbal tea bag	6
Raisins	42	Indian tea	250
Passion-fruit	39		
Sultanas	35	**Other**	
Prunes	27	Black treacle	140

Foods containing vitamin C (per 100 g/4 oz)

	mg		mg
Dairy		Watercress	60
Fresh whole milk	1.5	Cauliflower (boiled)	20
Natural yoghurt	0.4	Spring greens	30
		Avocado	15
Vegetables		Leeks (boiled)	15
Asparagus (boiled)	20	Lettuce	15
Runner beans (boiled)	5	Mustard and cress	40
Broad beans (boiled)	15	Onions (raw)	10
Broccoli florets (boiled)	34	Spring onions (raw)	25
Brussels sprouts (boiled)	40	Parsley	150
Cabbage, red (raw)	55	Parnips (boiled)	10
Radishes	25	Peas, fresh (boiled)	10
Spinach (boiled)	25		

Peppers, green (boiled)	60	Guavas (canned)	180
Potatoes (baked)	5–16	Lemons (whole)	80
		Lychees	40
Meat		Oranges	50
Lamb's kidney	9.0	Orange juice (fresh)	50
		Peaches (fresh)	8
Fruit (raw unless otherwise stated)		Pears (eating)	3
Apples	10	Pineapple (fresh)	25
Apples (baked with sugar)	14	Plums	3
Apricots (fresh)	7	Raspberries	25
Banana	10	Rhubarb (stewed)	8
Blackberries	20	Strawberries	60
Blackcurrants	200	Coconut (fresh)	2
Gooseberries (stewed)	31	Grapefruit juice (unsweet-	
Grapes (white)	4	ened)	28
Grapefruit	40		

Note: Nuts generally contain only a trace of vitamin C

Foods containing zinc (per 100 g/4 oz)

	mg		*mg*
Cereals		Lamb chops (lean only,	
Wheat bran	16.2	grilled)	4.1
Wholemeal (wholewheat)		Pork (lean only, grilled)	3.5
flour	3.0	Chicken (roast meat)	1.4
		Turkey (roast meat)	2.4
Dairy		Liver (fried)	6.0
Whole fresh milk	0.35		
Dried whole milk	3.2	**Nuts**	
Dried skimmed milk	4.1	Almonds	3.1
Cheddar cheese	4.0	Brazil nuts	4.2
Parmesan cheese	4.0	Hazelnuts	2.4
Yoghurt	0.60	Peanuts	3.0
		Walnuts	3.0
Eggs			
Eggs (boiled)	1.5	**Fish**	
Egg yolk (raw)	3.6	Cod (baked)	0.5
Egg (poached)	1.5	Plaice (fried)	0.7
		Herring (grilled)	0.5
Meat		Mackerel (fried)	0.5
Bacon (cooked)	0.8	Salmon (canned)	0.9
Beef (lean roast)	6.8	Sardines (canned in oil)	3.0
Lamb (cooked)	1.4		

	mg		mg
Tuna (canned)	0.8	Cabbage, red (raw)	0.3
Crab (boiled)	5.5	Lentils, split (boiled)	1.0
Prawns (boiled)	2.1	Peas, fresh (boiled)	0.5
Oysters (raw)	45.0	Lettuce	0.2
Mussels (boiled)	2.1	Spinach (boiled)	0.4
		Sweetcorn (boiled)	1.0
Vegetables and Pulses			
Butter beans (boiled)	1.0	**Other**	
Savoy cabbage (boiled)	0.2	Ginger (ground)	6.8

Foods containing iron (per 100 g/4 oz)

	mg		mg
Cereals		Mussels (boiled)	7.7
Bemax (wheat germ)	10.0	Oysters (raw)	6.0
Wheat bran	12.9	Scallops (steamed)	3.0
Rice	0.5		
		Nuts	
Meat		Almonds	4.2
Beef (lean cooked)	1.4	Brazil nuts	2.8
Rumpsteak (boneless sirloin) (lean, grilled)	3.5	Coconut (fresh)	2.1
Lamb (lean, roast)	2.5	**Dairy**	
Lamb kidney	12.0	Cheddar cheese	0.40
Pork (lean, grilled)	1.2		
Pig liver (stewed)	17.0	**Vegetables and Pulses**	
Veal	1.2	Haricot beans (boiled)	2.5
Chicken (dark meat)	1.0	Mung beans (raw)	8.0
Chicken liver (fried)	9.1	Red kidney beans	6.7
Bovril	14.0	Avocado	1.5
		Lentils (boiled)	2.4
Eggs		Butter beans (boiled)	1.7
Eggs, whole (boiled)	2.0	Parsley	8.0
Egg yolk (raw)	6.1	Spring greens (boiled)	1.3
		Leeks (boiled)	2.0
Fish			
Mackerel (fried)	1.2	**Fruit**	
Sardines (canned in oil)	2.9	Apricots (fresh)	0.4
Trout (steamed)	1.0	Bananas	0.4
Crab (boiled)	1.3	Blackberries	0.9
Prawns (boiled)	1.1	Dates (dried)	1.6
Cockles (boiled)	26.0	Figs (dried)	4.2

	mg		mg
Sultanas (dried)	1.8	Raisins	1.6
Prunes	2.9	Strawberries	0.7

Foods containing calcium (per 100 g/4 oz)

	mg		mg
Cereals		Rhubarb (stewed)	93
Brown flour	150	Tangerines	42
Oatmeal (uncooked)	55		
Soya flour	210	**Dairy**	
Wholemeal (wholewheat)		Milk	120
bread	23	Milk, dried (skimmed)	1190
Brown bread	100	Cheddar cheese	800
Muesli	200	Parmesan cheese	1220
		Cottage cheese	800
Fish		Yoghurt (natural)	180
Haddock (fried)	110		
Pilchards (canned in		**Vegetables and Pulses**	
tomato sauce)	300	Carrot (raw)	48
Sardines (canned in oil)	7	Celery (raw)	52
Sprats (fried)	710	Parsley (raw)	330
Tuna (canned in oil)	7	Spinach (boiled)	600
Shrimps (boiled)	320	Watercress	220
Whitebait (fried)	860	Turnips (boiled)	55
Salmon (canned)	93	French beans (boiled)	39
Kipper (baked)	65	Haricot beans (boiled)	65
Place (steamed)	38	Broccoli florets (boiled)	61
		Spring greens (boiled)	86
Fruit			
Apricots (dried)	92	**Nuts**	
Blackberries (raw)	63	Almonds	250
Figs (dried)	280	Brazil nuts	180
Lemons (whole)	110	Peanuts	61

Foods containing vitamin E (per 100 g/4 oz)

	mg		mg
Oils			
Cod liver oil	20.0	Peanut oil	13.0
Sunflower oil	48.7	Olive oil	5.1

	mg		mg
Meat		**Nuts**	
Lamb (cooked)	0.18	Almonds	20.0
Lamb kidney	0.41	Brazil nuts	6.5
Pork (cooked)	0.12	Hazelnuts	21.0
Chicken (roast meat)	0.11	Peanuts	8.1
Eggs		**Fruit**	
Eggs (boiled/poached)	1.6	Blackberries (raw)	3.5
Fish		Blackcurrants	1.0
Cod (baked)	0.59	**Vegetables**	
Halibut (grilled)	0.90	Asparagus (boiled)	2.5
Herring (grilled)	0.30	Broccoli florets (boiled)	1.1
Mussels (boiled)	1.2	Brussels sprouts (boiled)	0.9
Salmon (canned)	1.5	Parsley	1.8
Tuna (canned in oil)	6.3	Spinach (boiled)	2.0
		Avocado	3.2

Foods containing polyunsaturated fats

Certain fish contain essential oils, similar to those found in vegetables. These oils are helpful in maintaining skin quality and may also be of value in preventing pre-menstrual breast tenderness.

Herring	Sardines
Mackerel	Sprats
Pilchard	Whitebait
Salmon	

15

The 30-Day Diet

———— DAY 1 ————

Breakfast

Oats (soaked overnight) with chopped mixed fresh fruit
and milk

Mid-morning snack

Lunch

Cheese omelette
Brown Rice and Watercress Salad*

Mid-afternoon snack

Dinner

Grilled chicken
Orange Sauce*
Grilled mushrooms
Fresh greens
Potatoes

Dessert

Cinnamon Rhubarb*

Sucrose is the proper term for table sugar, most of
which is made from either sugar-cane or sugar-beet.

Brown Rice and Watercress Salad

serves 4

50 g (2 oz) brown rice, cooked
1 bunch of watercress, washed and chopped
100 g (4 oz) canned or frozen sweetcorn, cooked and
drained
1 green pepper, de-seeded and chopped
black pepper, to taste

Mix all the ingredients together and season with black pepper.

Orange Sauce

serves 4

100 ml (4 fl oz) white vinegar
1 tablespoon sugar
225 ml (8 fl oz) fresh orange juice, strained
1 teaspoon cornflour
1 tablespoon water
knob of butter

1 Put the vinegar and sugar in a saucepan and stir over a medium heat until the sugar has dissolved. Boil rapidly until the mixture turns a light golden brown.
2 Stir the orange juice into the saucepan and bring back to the boil. Reduce the heat and simmer until the liquid has reduced by about half.
3 Blend the cornflour with the water and add to the saucepan. Cook over a low heat, stirring constantly, until the mixture thickens.
4 Add the butter and mix in well before serving.

Cinnamon Rhubarb

serves 4

300 g (11 oz) rhubarb
4 tablespoons water
pinch of ground cinnamon
40–50 g (1½–2 oz) granulated or muscovado sugar

1 Wash, trim and chop the rhubarb.
2 Put the rhubarb in a saucepan with the water, cinnamon and sugar, and stew until the rhubarb is tender.
3 Spoon into separate dishes and serve.

Alternative Snacks

- Rice cakes. You can have sweet or savoury toppings, eg. sliced bananas, cheese, sugar-free peanut butter or jam.
- Rice salad. Add any combination of fruit, nuts, seeds, salad or vegetables.
- Potato salad. Add chives, onions or vegetables of your own choice with a small amount of low-calorie mayonnaise if required.
- Unsalted nuts or seeds, eg. peanuts, cashews, almonds, sunflower seeds, pumpkin seeds, honey-roasted sunflower seeds.
- Dried fruit and nut mix, eg. dried bananas, coconut, pineapple, apple, etc.
- Fresh fruit.
- Sugar-free fruit and nut bars.
- Ryvitas (with toppings as suggested for rice cakes).
- Low-fat or live yoghurt with added seeds, fruit or nuts.
- Raw carrots and celery.
- Savoury snacks, eg. Jordans 'Oatsters'.

DAY 2

Breakfast
Rice Krispies, milk and chopped nuts

Mid-morning snack

Lunch
Jacket potato with tuna
Choice of salad

Mid-afternoon snack

Dinner
Creamy Vegetable Bake*
Sauté potatoes

Dessert
Baked Bananas*

Creamy Vegetable Bake

serves 4

175 g (6 oz) carrots, diced
300 g (11 oz) broccoli florets
1 red pepper, de-seeded and diced
1 small leek, finely sliced
40 g (1½ oz) vegetable margarine
25 g (1 oz) plain flour
450 ml (¾ pint) milk
2 teaspoons creamed horseradish
black pepper, to taste
200 g (7 oz) canned or frozen sweetcorn
175 g (6 oz) dried breadcrumbs

1 Steam the carrots and broccoli for 6 minutes, then add the red pepper and leek and steam for a further 6 minutes.
2 Meanwhile, melt the margarine in a saucepan and stir in the flour. Cook, stirring, for 1 minute, then remove from the heat and gradually stir in the milk.
3 Bring to the boil, stirring constantly, then reduce the heat and simmer gently for 2 minutes or until smooth and thick. Stir in the horseradish and season with pepper.
4 Place all the cooked vegetables and the sweetcorn in an ovenproof dish, pour over the sauce and sprinkle with the breadcrumbs.
5 Bake in a preheated oven at 180°C (350°F) mark 4 for 10 minutes. Serve with potatoes.

Baked Bananas

serves 6

6 large bananas
1 tablespoon vegetable oil

1 Rub the banana skins with a little of the oil. Brush an ovenproof dish with the remaining oil.
2 Lay the bananas in the oiled dish and bake in the centre of a preheated oven at 170°C (325°F) mark 3 for 30–40 minutes or until the bananas are soft and the skins have turned black.
3 Serve the bananas hot, with one strip of the skin peeled back.

There are many different types of sugars, including glucose, fructose, sucrose and lactose (milk sugar), amongst others.

DAY 3

Breakfast
Scrambled egg with grilled tomatoes
Toast or rice cakes with sugar-free jam

Mid-morning snack

Lunch
Cauliflower Soup*
Tropical Rice Salad*

Mid-afternoon snack

Dinner
Chilli Haddock Casserole*
Red cabbage
Jacket potato

Dessert
Apricot Sherbet*

Cauliflower Soup

serves 4

50 g (2 oz) sunflower margarine
100 g (4 oz) plain flour
300 ml (½ pint) milk
900 ml (1½ pints) Home-made Chicken Stock (see opposite)
1 large cauliflower, stalk removed and broken into florets
1 teaspoon dried chervil
¼ teaspoon ground mace
black pepper, to taste
1 tablespoon single cream

1 Melt the margarine in a large saucepan. Add the flour and cook for 1 minute, stirring constantly. Remove from the heat and gradually stir in the milk and stock until the mixture is smooth.

2 Add the cauliflower, chervil, mace and black pepper and simmer gently for 15 minutes or until the cauliflower is just soft. Mash the cauliflower well or purée in a blender or food processor.

3 Stir in the single cream and serve.

Home-made Chicken Stock

makes 600 ml (1 pint)

900 ml (1½ pints) water
chicken bones, cooked (ie. left over from roast chicken) or uncooked
100 g (4 oz) mixed carrots, celery and onions, chopped
chopped fresh herbs
black pepper, to taste

1 Put the water, chicken bones, vegetables and herbs in a large saucepan and bring to the boil. Season with pepper and simmer for 1–2 hours, then strain into a bowl and leave to cool. When cool, put in the refrigerator and chill for 1 hour.

2 Once the stock is chilled, remove it from the refrigerator and remove any fat from the top. The chicken stock is now ready to use or to freeze.

All sugars have a sweet taste but differ in their chemical structure. Glucose and fructose are composed of single sugar molecules and are termed monosaccharides.

Tropical Rice Salad

serves 4–6

175 g (6 oz) American long-grain rice
½ teaspoon ground turmeric
450 ml (¾ pint) vegetable stock
2 ripe bananas
2 tablespoons lemon juice
½ pineapple, peeled, cored and chopped
100 g (4 oz) sultanas
½ cucumber, cubed

1 Put the rice, turmeric and vegetable stock in a saucepan and bring to the boil. Reduce the heat and simmer for 10–15 minutes. Drain thoroughly and allow to cool.
2 Peel and slice the bananas and toss in the lemon juice.
3 Mix the rice, banana, pineapple, sultanas and cucumber together and put in a salad bowl.

Chilli Haddock Casserole

serves 4

25 g (1 oz) sunflower margarine
4 courgettes, sliced
1 onion, sliced
1 small fresh chilli, de-seeded and finely chopped
4 tomatoes, skinned and chopped or 230 g (8.1 oz) can chopped tomatoes
750 g (1½ lb) haddock fillet, skinned and cut into chunks
black pepper, to taste
1 tablespoon sesame seeds

1 Melt the margarine in a frying pan. Add the courgettes and onion and fry gently for 5 minutes. Add the chilli and tomatoes and cook for a further 3–4 minutes.

2 Put the fish in an ovenproof dish, cover with the vegetable mixture and season with black pepper. Cover and bake in a preheated oven at 200°C (400°F) mark 6 for 15 minutes, then stir in the sesame seeds. Bake for a further 5 minutes, then serve.

Variation
Top the casserole with mashed potato or a sprinkling of grated cheese and brown under the grill.

Apricot Sherbet
serves 4–6

350 g (12 oz) fresh apricots, halved
150 ml (¼ pint) water
100 g (4 oz) sugar
1 teaspoon lemon juice
300 ml (½ pint) double cream

1 Put the apricots, with their stones, in a saucepan with the water, sugar and lemon juice. Cook gently until the apricots are soft, stirring occasionally. Remove the stones.
2 Transfer the apricots and liquid to a blender or food processor and blend to a smooth purée. Pour 300 ml (½ pint) of the purée into a measuring jug and leave to cool.
3 Whip the cream until it holds its shape, then gradually add the apricot purée, whipping constantly until evenly mixed.
4 Pour the mixture into a freezer container and freeze for at least 4 hours, preferably overnight.
5 Before serving, transfer the sherbet to the refrigerator for about 10 minutes to soften slightly, then serve in individual dessert glasses.

DAY 4

Breakfast

Low-fat yoghurt with chopped fruit and nuts

Mid-morning snack

Lunch

Cottage cheese and cucumber sandwich

Mid-afternoon snack

Dinner

Lamb Cutlets with Nutty Apricot and Mint Stuffing*
Spinach
Sweetcorn
New potatoes

Dessert

Blackcurrant Sorbet*

Sucrose and lactose are composed of two single sugar molecules joined together and are thus termed disaccharides. Sucrose is composed of one part of glucose joined to one part of fructose.

Lamb Cutlets with Nutty Apricot and Mint Stuffing

serves 4

3 dried (or fresh) apricots
1½ tablespoons sunflower oil
1 small onion, finely chopped
100 g (4 oz) chopped mixed nuts
25 g (1 oz) fresh breadcrumbs
1 tablespoon vegetable stock
1 tablespoon chopped fresh mint
8 lamb cutlets, trimmed of most excess fat

1 If using dried apricots, put them in a bowl, cover with water and leave to soak for 1 hour, then drain. Chop the apricots. Stone and chop fresh apricots, if using.
2 Heat 1 tablespoon of the oil in a frying pan and gently fry the onion until transparent.
3 Mix the nuts, breadcrumbs, stock, mint, apricots and onion together. Put 1 tablespoon of the mixture in the hollow of each cutlet, making a round shape of the stuffing and chop together.
4 Brush the cutlets with the remaining oil and cook under a medium grill for 8–10 minutes on each side or until the meat is tender, carefully turning once with a fish slice.

Blackcurrant Sorbet

serves 4–6

300 ml (½ pint) water
100 g (4 oz) granulated sugar
juice of ½ lemon
450 g (1 lb) fresh or frozen blackcurrants
2 egg whites

1 Put the water and sugar in a heavy-based saucepan and heat gently until the sugar has dissolved. Bring to the boil, then boil for 5 minutes or until syrupy. Remove from the heat, stir in the lemon juice and leave to cool.

2 Put the blackcurrants in a heavy-based saucepan and heat gently for 5–10 minutes or until the blackcurrants are soft and the juices run. Remove from the heat and press the blackcurrants through a sieve. Leave to cool.

3 Combine the syrup and blackcurrant purée, pour the mixture into a freezer container, and chill in the refrigerator for at least 30 minutes.

4 Transfer the sorbet to the freezer and freeze for 1–2 hours or until slushy.

5 Whisk the egg whites until stiff. Remove the blackcurrant mixture from the freezer and turn it into a bowl. Beat thoroughly, then fold in the egg whites until evenly blended.

6 Return the sorbet mixture to the container, cover and freeze for at least 2 hours or until firm. (The sorbet can be stored in the freezer for up to 2 months.)

7 Before serving, stand the sorbet at room temperature for 10–20 minutes to soften slightly. Scoop into individual bowls.

DAY 5

Breakfast

Poached egg on toast
Grilled tomatoes

Mid-morning snack

Lunch

Lean ham
Bean Sprout Salad*
Potato Salad*

Mid-afternoon snack

Dinner

Chicken with Peanut Sauce*
Brown rice
Broccoli

Dessert

Baked Apple*

Bean Sprout Salad

serves 4

175 g (6 oz) bean sprouts
100 g (4 oz) red pepper, de-seeded and sliced
100 g (4 oz) carrots, grated
4 spring onions, chopped
salad dressing (see overleaf)

Mix all the ingredients together and toss in the salad dressing of your choice.

Mustard Dressing

100 ml (4 fl oz) lemon juice
1 tablespoon French whole grain mustard
1 tablespoon chopped fresh parsley
3 tablespoons water

Whisk all the ingredients together until well combined.
Cover and chill before serving.

Coconut Mayonnaise

3 tablespoons coconut milk
3 tablespoons low-calorie mayonnaise
1 tablespoon lemon juice
½ teaspoon caster sugar
1 teaspoon curry powder

Whisk all the ingredients together until well combined.
Cover and chill before serving.

Orange Chilli Dressing

2 tablespoons light soy sauce
3 tablespoons orange juice
1 small fresh green chilli, de-seeded and chopped
1 garlic clove, crushed

Whisk all the ingredients together until well combined.
Cover and chill before serving.

Potato Salad

serves 4–6

700 g (1½ lb) cooked potatoes
1 green pepper, de-seeded and sliced
3 tablespoons snipped fresh chives
6 tablespoons low-calorie mayonnaise
black pepper, to taste

Chop the potatoes into bite-sized pieces. Mix all the ingredients together and serve.

Chicken with Peanut Sauce

serves 4

2 tablespoons vegetable oil
450 g (1 lb) skinless chicken breast fillets, cut into thin strips
1 teaspoon chilli powder
227 g (8 oz) can tomatoes, finely chopped
2 tablespoons crunchy peanut butter
1–2 tablespoons chilli sauce

1 Heat the oil in a wok or frying pan. Add the chicken and stir-fry for 2–3 minutes, sprinkling with the chilli powder while cooking. Remove from the pan with a slotted spoon and set aside.
2 Put the tomatoes, peanut butter and chilli sauce into the pan and stir well.
3 Return the chicken to the pan and bring to the boil, then reduce the heat and simmer gently, stirring occasionally, until the chicken is tender. Serve immediately.

Baked Apple

serves 1

1 cooking apple
2 teaspoons concentrated apple juice
250 ml (8 fl oz) water
pinch of ground cinnamon

1 Wash and core the apple. Cut around the centre of the apple, just scoring the skin. Put the apple in an ovenproof dish.
2 Mix the concentrated apple juice with the cinnamon.
3 Pour the water into the dish around the apple, and pour the apple juice over the apple.
4 Bake in a preheated oven at 180°C (350°F) mark 4 for 50–60 minutes or until the apple is soft.

All the sugars are a good source of energy. However, they require vitamin B, magnesium and other nutrients for the energy to be released. Unfortunately, many sources of sugars are poor in these essential nutrients.

DAY 6

Breakfast
Corn Flakes with milk and 1 tablespoon of linseeds

Mid-morning snack

Lunch
Smoked Mackerel Pâté*
Ryvita
Green salad

Mid-afternoon snack

Dinner
Beef Stir-Fry with Apricots and Walnuts*
Green pasta

Dessert
Chopped fresh fruit of choice and yoghurt

Smoked Mackerel Pâté

serves 4

50 g (2 oz) vegetable margarine
225 g (8 oz) smoked mackerel fillets, skinned and any
bones removed
juice of ½ orange
1 teaspoon tomato purée
1 teaspoon white wine vinegar
black pepper, to taste

1 Beat the margarine until very soft, then put in a blender
or food processor with the remaining ingredients. Purée
until smooth.
2 Spoon the pâté into four individual dishes or ramekins
and chill before serving with pitta bread or crudités.

Beef Stir-fry with Apricots and Walnuts

serves 4

6 fresh or dried apricots
350 g (12 oz) rump steak
2 teaspoons cornflour
4 tablespoons water
4 tablespoons orange juice
2 teaspoons groundnut (peanut) or vegetable oil
4 spring onions, cut diagonally into 2.5 cm (1 inch) pieces
1 tablespoon Worcestershire sauce
3 medium Chinese leaves, roughly chopped
50 g (2 oz) walnut pieces
black pepper, to taste

1 If using dried apricots, soak them in cold water for 1 hour, then drain and cut them into quarters. If using fresh apricots, stone them and cut them into quarters.
2 Remove any fat from the steak, wrap the meat in cling film and place it in the freezer for about 45 minutes or until nearly frozen.
3 Remove the meat from the freezer and cut it across the grain into very thin strips.
4 Blend the cornflour with 1 tablespoon of water, then add the rest of the water and the orange juice.
5 Heat the oil in a wok or frying pan. Add the meat and stir-fry for 3–4 minutes or until browned. Reduce the heat to moderate, add the onions, Worcestershire sauce and Chinese leaves, and continue to stir-fry for a further minute.
6 Add the apricots and cornflour mixture to the pan and bring to the boil over a moderate to high heat, stirring constantly. After about 30 seconds, the mixture should become thick and glossy.
7 Remove the pan from the heat and stir in the walnuts. Season with pepper and serve immediately.

Fruits naturally contain a variety of sugars, particularly fructose but also some sucrose. In the production of alcoholic beverages, these and other sugars are broken down by yeast into alcohol and carbon dioxide. The longer the fermentation process, the more sugar is broken down and the higher the alcohol content. Strong beers, for example, contain more alcohol but less sugar. The calorie content of beverages reduces only slightly with increased fermentation.

DAY 7

Breakfast
2 slices of toast with sugar-free jam

Mid-morning snack

Lunch
Jacket potato, 50 g (2 oz) grated cheese
Apple and Nut Salad*

Mid-afternoon snack

Dinner
Prawn-stuffed Tomatoes*
Brown rice
Courgettes
Green beans

Dessert
Pineapple Water-Ice*

Apple and Nut Salad
serves 4

4 red apples, wiped
lemon juice, to coat
½ a cucumber, thickly sliced
6 celery sticks, chopped
1 bunch of spring onions, sliced
75 g (3 oz) natural peanuts
dressing (see page 118)

1 Core and roughly chop the apples and dip them in the lemon juice to prevent discolouration. Cut the cucumber slices into quarters.
2 Mix all the ingredients together and toss in the dressing of your choice (see page 118).

Prawn-stuffed Tomatoes

serves 4

4 large tomatoes
50 g (2 oz) soup pasta (pastina)
175 g (6 oz) peeled cooked prawns
grated rind of 1 lemon
1 teaspoon paprika
1 teaspoon tomato purée
black pepper, to taste
25 g (1 oz) low-fat cheese, grated

1 Prepare the tomatoes by cutting off the tops and scooping out the insides. Put the tomato seeds and pulp in a bowl and set aside.
2 Cook the pasta in boiling water for about 5 minutes or until tender. Drain well.
3 Add the prawns, lemon rind, paprika, tomato purée and pepper to the tomato pulp. Add the cooked pasta and mix well.
4 Fill the tomatoes with the prawn mixture and top with a light sprinkling of cheese.
5 Stand the tomatoes in a baking dish and bake in a pre-heated oven at 180°C (350°F) mark 4 for 10 minutes. Serve immediately.

In practice, we consume sugars in two forms – those that occur naturally in fruits, vegetables and milk, and refined sugars that are normally added to foods, or used in beverages.

Pineapple Water-Ice

serves 4–6

450 ml (¾ pint) water
100 g (4 oz) granulated sugar
1 strip of lemon rind
1 fresh, ripe pineapple, weighing about 1.1 kg (2½ lb)

1 Put the water, sugar and lemon rind in a heavy-bottomed saucepan and heat gently until the sugar has dissolved.
2 Bring to the boil, then boil for 5 minutes or until syrupy. Remove from the heat and leave to cool. Discard the lemon rind.
3 Cut the pineapple in half lengthways and scoop out the flesh. Wrap the shells closely in foil and chill in the refrigerator.
4 Purée the pineapple flesh in a blender or food processor and put 450 ml (¾ pint) of it in a measuring jug.
5 Combine the syrup and pineapple purée, then pour into a freezer container, cover and chill in the refrigerator for at least 30 minutes. Transfer to the freezer for 1–2 hours or until slushy.
6 Remove the mixture from the freezer and turn it into a bowl. Beat thoroughly, then return to the freezer container. Cover and freeze for at least 2 hours or until firm. (The sorbet can be stored in the freezer for up to 2 months.)
7 Before serving, stand the sorbet at room temperature for 10–20 minutes to soften slightly, then scoop it into the chilled pineapple shells. Serve immediately.

DAY 8

Breakfast
Boiled egg
Toast

Mid-morning snack

Lunch
Baked Avocado with Tuna*
Potato Salad* (see page 119)
Endive, Fruit and Nut Salad

Mid-afternoon snack

Dinner
Broccoli with Coconut and Cashews*
Brown rice
Green beans

Dessert
Yoghurt with chopped banana

Baked Avocado with Tuna

serves 4

25 g (1 oz) vegetable margarine
50 g (2 oz) plain flour
150 ml (¼ pint) semi-skimmed milk
black pepper, to taste
99 g (3½ oz) can tuna in oil, well drained and flaked
1 tablespoon lemon juice
2 large ripe avocados
25 g (1 oz) Gruyère cheese, grated
lemon slices, to garnish

1 Melt the margarine in a saucepan, stir in the flour and cook for 1½ minutes, stirring constantly.

2 Remove from the heat and gradually stir in the milk. Bring slowly to the boil, stirring constantly, then simmer for 2 minutes, stirring, until thickened. Season with pepper and remove from the heat.

3 Stir the tuna and lemon juice into the sauce.

4 Cut the avocados in half and remove the stones. Stand the avocado halves in a baking dish, using crumpled foil, if necessary, to help them stand upright.

5 Spoon the tuna mixture on to the avocados, covering all the avocado flesh. Sprinkle with grated cheese and bake in a preheated oven at 180°C (350°F) mark 4 for 15–20 minutes.

6 Transfer the avocados to serving dishes, garnish with lemon slices and serve immediately.

Endive, Fruit and Nut Salad

serves 4 as a main course; 8 as a starter or side dish

2 heads curly endive (frisée)
3 oranges
25 g (1 oz) flaked almonds
25 g (1 oz) walnuts, chopped
2 apples, washed, cored and sliced into small segments
75 g (3 oz) seedless grapes
1 tablespoon lemon juice

1 Pull the endive apart, wash and dry it thoroughly. Tear the leaves into pieces and place in a salad bowl.

2 Grate the rind of one orange into a bowl. Remove the peel and pith from all the oranges, break them into segments and place in a separate bowl.

3 Mix together the almonds, walnuts, apple, grapes and lemon juice, and add to the oranges. Mix well.

4 Place the fruit and nut mixture in the bowl on top of the endive. Chill before serving with a dressing of your choice (see page 118).

Broccoli with Coconut and Cashews

serves 4

200 g (7 oz) creamed coconut
300 ml (½ pint) boiling water
275 g (10 oz) broccoli, stalks sliced and tops broken into
florets
1 tablespoon vegetable oil
1 medium onion, finely chopped
2.5 cm (1 inch) piece of fresh root ginger, peeled and
finely chopped
75 g (3 oz) cashew nuts
¼ teaspoon ground turmeric
2 teaspoons ground coriander
black pepper, to taste

1 Put the coconut in a bowl, pour over the boiling water
and stir until dissolved.
2 Steam or parboil the broccoli for 2–3 minutes, then rinse
under cold running water and drain.
3 Heat the oil in a wok or frying pan until hot. Add the
onion, ginger, cashews, turmeric and coriander and stir-fry
gently for 2–3 minutes, without browning.
4 Add the coconut milk and pepper to taste, then simmer
gently for 2–3 minutes or until the mixture thickens.
5 Add the broccoli and simmer for a further 2 minutes.
Serve immediately.

It's the 'added sugars' that are strongly associated
with dental caries (tooth decay). This includes
sucrose as well as glucose and fructose. Dental caries
is one of the prices we have had to pay for the
refining and processing of foods.

DAY 9

Breakfast

Rice Krispies with 1 tablespoon seeds

Mid-morning snack

Lunch

Potato and Basil Soup*
Jacket potato with sweetcorn

Mid-afternoon snack

Dinner

Spanish Chicken*
Brown rice
Greens

Dessert

Yoghurt with Nut de la Crème*

Potato and Basil Soup

serves 4

1 tablespoon sunflower oil
700 g (1½ lb) potatoes, peeled and grated
3 garlic cloves, crushed
50 g (2 oz) fresh basil, chopped
900 ml (1½ pints) chicken stock (see page 111)
pinch of low-sodium salt
black pepper, to taste

1 Heat the oil in a large saucepan and gently fry the potatoes, garlic and half of the basil for 2 minutes, then add the chicken stock.

2 Bring to the boil, then reduce the heat, cover and simmer for 15–20 minutes or until the potatoes are soft.
3 Season with salt and pepper and mash well or purée in a blender or food processor. Add the remaining basil and serve hot with herb bread.

Spanish Chicken

serves 6

2 tablespoons sunflower oil
6 chicken portions
1 large onion, finely choped
1 green pepper, de-seeded and chopped
1 red pepper, de-seeded and chopped
300 ml (½ pint) chicken stock (see page 111)
225 g (8 oz) mushrooms, sliced
6 tomatoes, sliced
10 stuffed green olives, cut in half
10 pitted black olives

1 Heat half the oil in a frying pan and fry the chicken pieces until lightly browned all over. Place the chicken in a casserole and set aside.
2 Add the onion and peppers to the oil remaining in the frying pan and fry gently for 5 minutes. Remove with a slotted spoon and spoon over the chicken in the casserole.
3 Pour the stock into a saucepan and bring to the boil. Pour into the casserole, cover and bake in a preheated oven at 180°C (350°F) mark 4 for 1–1½ hours or until the chicken is tender.
4 Ten minutes before the end of the cooking time, heat the remaining oil in a frying pan and gently fry the mushrooms and tomatoes for about 4 minutes.
5 Add the mushrooms, tomatoes and olives to the casserole and serve immediately.

Nut de la Crème

serves 2

50 g (2 oz) mixed nuts

1 Grind the nuts finely, using the fine attachment on a mincer or a coffee grinder.
2 Mix the ground nuts with enough cold water to form a cream. Use as a cream on fruit, yoghurt, salads or vegetables.

Variations
By using just one type of nut or a combination of a few different types of nuts, the flavour of the cream can be varied.

In theory, the human body could exist on a diet completely devoid of all types of sugar, surviving simply on protein and fats as a source of calories and other nutrients. Accordingly, there is no set dietary requirement for naturally occurring sugars in food, or for added sugars.

DAY 10

Breakfast
Yoghurt with chopped fresh fruit and nuts

Mid-morning snack

Lunch
Stir-fried vegetables with brown rice

Mid-afternoon snack

Dinner
Liver and Bacon in Tomato Sauce*
Potatoes
Spinach
Carrots

Dessert
Fruit Snow*

Liver and Bacon in Tomato Sauce
serves 4

1 tablespoon sunflower oil
1 large onion, sliced
450 g (1 lb) lamb's liver, sliced
50 g (2 oz) plain flour seasoned with black pepper
4 rashers of lean unsmoked back bacon, de-rinded
397 g (14 oz) can tomatoes
1 tablespoon cornflour
3 tablespoons water

1 Heat the oil in a frying pan and gently fry the onion until transparent. Remove the onion with a slotted spoon and place it in a casserole.
2 Toss the liver in the seasoned flour. Add the liver to the oil remaining in the frying pan and cook until lightly

browned. Remove the liver from the pan and place it on top of the onion in the casserole.

3 Place the bacon rashers on top of the liver. Add the tomatoes with their juice, cover and cook in a preheated oven at 180°C (350°F) mark 4 for 30 minutes.

4 Five minutes before the end of the cooking time, dissolve the cornflour in the water and stir into the casserole. Return the casserole to the oven for the final 5 minutes' cooking.

Fruit Snow

serves 2

200 g (7 oz) dessert apples, peeled, cored and thinly sliced
2 tablespoons water
1 teaspoon grated orange rind
orange slices, to decorate

1 Put the apple, water and orange rind in a saucepan, cover and cook gently, stirring occasionally, until the apples are soft.

2 Press the apple mixture through a sieve and leave to cool. Alternatively, purée in a blender or food processor.

3 Whisk the egg white until stiff and fold into the apple. Spoon the mixture into two dessert glasses and chill before serving. Serve decorated with slices of orange.

The brain and nervous system has an absolute requirement for glucose. Normally, this is derived from sugars in the diet but can also be produced by processing fats and carbohydrates which the liver can turn into glucose.

DAY 11

Breakfast

Oats (soaked overnight) with milk, nuts and seeds

Mid-morning snack

Lunch

Seafood Salad*

Pasta Salad*

Mid-afternoon snack

Dinner

Baked Chicken Burgers*

Sauté potatoes

Peas

Broccoli

Dessert

Stewed fruit with Frumble Topping*

In practice, our diets contain not only natural sugars in fruit and vegetables, and added sugars, but also carbohydrates. These are complex compounds composed of multiple sugar units joined together; they form the bulk of the structure of many fruits and vegetables. They themselves do not taste sweet, but when broken down by the body's digestive processes, they release sugars.

Seafood Salad

serves 4

350 g (12 oz) cod fillet
½ white cabbage, trimmed
225 g (8 oz) peeled prawns
100 g (4 oz) cooked mussels, shelled
1 onion, grated
100 g (4 oz) carrots, grated
½ teaspoon chopped fresh dill
dressing or mayonnaise (see page 118)

1 Steam the cod until tender, then remove the skin and flake the fish. Leave to cool.
2 Shred the cabbage finely, then rinse and drain well. Pat dry with a clean cloth or kitchen paper.
3 Mix the cod, cabbage, prawns, mussels, onion, carrots and dill together.
4 Toss the salad in the dressing of your choice (see page 118) or in low-calorie mayonnaise.

Pasta Salad

serves 4

225 g (8 oz) dried pasta shapes
1 teaspoon sunflower oil
1 red pepper, de-seeded and chopped
100 g (4 oz) raisins
175 g (6 oz) canned or frozen sweetcorn, cooked and
drained
3 tablespoons sunflower oil
2 tablespoons lemon juice
1 garlic clove, crushed
1 tablespoon chopped fresh parsley

1 Cook the pasta in boiling water, with 1 teaspoon oil added, for about 10 minutes or until *al dente* (tender but still slightly firm to the bite). Drain the pasta and rinse under cold running water.

2 Mix the pasta with the pepper, raisins and sweetcorn.

3 Beat together the remaining 3 tablespoons sunflower oil, the lemon juice, garlic and parsley, and pour over the pasta. Toss to coat thoroughly.

Baked Chicken Burgers

serves 4

450 g (1 lb) cooked chicken meat, minced or finely chopped
1 egg (size 3–4), beaten
1 garlic clove, crushed
1 tablespoon finely chopped fresh parsley
2 teaspoons dried tarragon
black pepper, to taste
1 tablespoon sunflower oil
1 onion, finely chopped

1 Mix the chicken and egg together. Add the garlic, parsley, tarragon and pepper, and mix well.

2 Heat the oil in a frying pan and gently fry the onion for 2–3 minutes. Add the onion to the chicken mixture and stir well.

3 Form the mixture into round burger shapes and place on a greased baking tray.

4 Bake in a preheated oven at 180°C (350°F) mark 4 for 15–20 minutes or until golden brown, turning once halfway through baking.

Frumble Topping

serves 4

4 slices of white bread, crusts removed
50 g (2 oz) butter or margarine

1 Crumble the bread to coarse crumbs, using your hands.
2 Melt the butter or margarine in a frying pan and sprinkle in the crumbs. Fry until crisp and golden, turning the crumbs continuously with a spatula.
3 Serve as a topping for hot stewed fruit, with a little sugar sprinkled on top, if required.

DAY 12

Breakfast
2 slices of toast or rice cakes with sugar-free jam

Mid-morning snack

Lunch
Jacket potato
Red Lentil Dal*
Salad of choice

Mid-afternoon snack

Dinner
Trout with Watercress, Grape and Yoghurt Dressing*
Potatoes
Broad beans
Carrots

Dessert
Rhubarb and Ginger Mousse*

Red Lentil Dal

serves 4

225 g (8 oz) split red lentils
1 tablespoon sunflower oil
1 medium onion, finely chopped
2 garlic cloves, crushed
1 teaspoon paprika
1 teaspoon ground cumin
½ green chilli, de-seeded and chopped
3 ripe tomatoes, skinned and chopped
1 tablespoon lemon juice
black pepper, to taste

1 Wash the lentils well and place in a pan of water. Bring to the boil, then reduce the heat and simmer for 15 minutes. Drain well.

2 Heat the oil in a frying pan and gently fry the onion and garlic for 2–3 minutes.

3 Add the paprika, cumin and chilli, and cook for a further 1 minute. Stir in the tomatoes, lentils and lemon juice, and season with black pepper. Heat through and serve.

Trout with Watercress, Grape and Yoghurt Dressing

serves 2

2 trout, cleaned
2 tablespoons water
1 tablespoon skimmed milk
½ bunch of watercress, roughly chopped
100 g (4 oz) seedless grapes, halved
2 teaspoons lemon juice
6 tablespoons natural low-fat yoghurt

1 Place each trout on a large square of foil. Add 1 tablespoon water to each, wrap in the foil and seal. Bake in a preheated oven at 180°C (350°F) mark 4 for 20–25 minutes.

2 Add the milk, watercress, grapes and lemon juice to the yoghurt.

3 When the trout are cooked, remove them from the oven and leave to cool.

4 Chill the trout and the dressing in the refrigerator before serving.

Variation
The trout can also be served hot.

Rhubarb and Ginger Mousse
serves 4

450 g (1 lb) rhubarb
3 tablespoons clear honey
grated rind and juice of ½ orange
¼ teaspoon ground ginger
2 tablespoons water
2 teaspoons powdered gelatine
2 egg whites

1 Trim the rhubarb and chop it into 2.5 cm (1 inch) pieces.
2 Put the rhubarb in a saucepan with the honey, orange rind, orange juice, and ginger, and simmer gently until the fruit is soft.
3 Put the water in a small heatproof bowl and sprinkle in the gelatine. Leave to soften for 5 minutes, then stand the bowl in a saucepan of water and heat gently, stirring, until the gelatine has dissolved.
4 When the rhubarb is cooked, leave it to cool slightly, then purée the mixture in a blender or food processor until smooth. Stir in the gelatine and leave to cool until half set.
5 Whisk the egg whites until stiff and fold them lightly into the half-set rhubarb mixture. Spoon into four dessert glasses and chill until set.

Undigestible carbohydrates are also known as fibre, which has important health properties. Healthy eating is about achieving a correctly balanced dietary intake of fibre, fats, protein, sugars, vitamins and minerals.

DAY 13

Breakfast

Yoghurt with chopped banana and 1 tablespoon seeds

Mid-morning snack

Lunch

100 g (4 oz) turkey breast
Coleslaw Salad*
Potato Salad* (see page 119)

Mid-afternoon snack

Dinner

Roast lamb
Chestnut Sauce*
Green beans
Cabbage
Potatoes

Dessert

Gooseberry Crumble*

Coleslaw Salad

serves 6

1 medium white cabbage, cored and finely shredded
5 medium carrots, coarsely grated
1 medium onion, finely chopped
50 g (2 oz) raisins
8 tablespoons mayonnaise
black pepper, to taste

1 Mix the cabbage, carrots, onion and raisins in a large bowl.

2 Add the mayonnaise and pepper and toss gently to coat all the ingredients.

Chestnut Sauce

serves 4

225 g (8 oz) fresh chestnuts, peeled
300 ml (½ pint) chicken stock (see page 111)
½ small onion, finely chopped
1 small carrot, chopped
40 g (1½ oz) vegetable margarine
3 tablespoons plain flour
black pepper, to taste
2 tablespoons single cream

1 Put the chestnuts, stock and vegetables in a saucepan and simmer until soft. Purée in a blender or food processor, mash well, or press through a sieve.

2 Melt the margarine in a saucepan and stir in the flour. Add the chestnut purée and bring to the boil, stirring, until the sauce is thick. If the sauce seems too stiff, add a little milk.

3 Season the sauce with pepper, remove from the heat and stir in the cream. Re-heat, without boiling, before serving.

The average consumption of added sugars (mainly sucrose but also some glucose and fructose) is approximately 40 kg (88 lb) per person per year. This provides nearly 20 per cent of dietary energy. Intakes at this level are strongly associated with dental caries, which is relatively rare if the consumption is halved. The 1991 recommendation from the Department of Health was that current consumption of sucrose should be halved.

Gooseberry Crumble

serves 4–6

450 g (1 lb) gooseberries, topped and tailed
concentrated apple juice, to taste
100 g (4 oz) margarine
225 g (8 oz) plain flour
50 g (2 oz) sugar

1 Put the gooseberries in a saucepan with a little water. Sweeten with concentrated apple juice and spoon into an ovenproof dish.
2 Rub the margarine into the flour until the mixture resembles fine breadcrumbs. Stir in the sugar and sprinkle on top of the gooseberries.
3 Bake in a preheated oven at 230°C (450°F) mark 8 for 10–15 minutes. Serve hot or cold.

Variations
Any kind of fruit in season is suitable, such as apples, rhubarb, apricots, plums, blackcurrants, blackberry and apple, pears or any mixture.

DAY 14

Breakfast

2-egg omelette
Grilled tomatoes
Grilled mushrooms
Toast

Mid-morning snack

Lunch

Greek Salad*
100–150 g (4–5 oz) jacket potato

Mid-afternoon snack

Dinner

Fish Balls*
Light Cheese Sauce*
Peas
Broccoli
Brown rice

Dessert

Summer Pudding*

Unfortunately, in the short term, many people tolerate and indeed enjoy having a high intake of added sugar. In addition to the general population halving sucrose consumption, those who are overweight or diabetic, and others with certain metabolic problems, should take particular care to limit sucrose consumption severely.

Greek Salad

serves 4

½ cucumber, thickly sliced
1 small onion, sliced
4 large tomatoes, sliced
4 black olives
100 g (4 oz) feta cheese, drained and cut into squares
4 tablespoons olive oil
1 tablespoon lemon juice
2 teaspoons chopped fresh oregano
½ small iceberg lettuce, shredded

1 Cut the cucumber slices into quarters and put in a bowl with onion, tomatoes, olives and cheese. Mix well.
2 Beat the oil, lemon juice and oregano together until thoroughly combined.
3 Place the lettuce in a glass bowl and spoon the salad on top. Pour over the dressing.

Fish Balls

makes 30–34 fish balls

450 g (1 lb) white fish fillets, skinned
1 egg white
2 teaspoons cornflour
2 teaspoons vegetable oil
2 tablespoons finely chopped fresh coriander
black pepper, to taste
900 ml (1½ pints) fish stock

1 Put the fish in a blender or food processor and purée until smooth. Add the egg white, cornflour oil, coriander and pepper to taste. Blend again until thoroughly combined, then turn into a bowl.

2 Using lightly floured hands, roll the mixture into 30–34 small balls. Place on a floured board and refrigerate for 25–30 minutes.

3 Put the fish stock in a large saucepan and bring to the boil. Drop the fish balls into the stock and simmer gently for 5–6 minutes. This can be done in two batches. Remove from the stock with a slotted spoon and serve immediately with vegetables and a Light Cheese Sauce (see below), or serve cold with salad.

Variations
Instead of poaching the fish balls, try stir-frying them in oil flavoured with garlic for 4–5 minutes.
Substitute chilli or curry powder for the coriander.

Light Cheese Sauce
makes 450 ml (¾ pint)

25 g (1 oz) vegetable margarine
2 tablespoons plain flour
450 ml (¾ pint) semi-skimmed milk
100 g (4 oz) low-fat cheese, grated
pepper, to taste

1 Melt the margarine in a saucepan. Add the flour and cook over a low heat, stirring, for 1 minute.

2 Remove from the heat and gradually stir in the milk. Return to the heat and slowly bring back to the boil, stirring constantly. Simmer for a further 2 minutes or until the sauce is thick and smooth.

3 Stir in the cheese and season with pepper. Heat gently without boiling.

Variation
Mustard can be added to the sauce to add to the flavour.

Summer Pudding

serves 4

450 g (1 lb) blackcurrants, redcurrants and raspberries
4 tablespoons water
2–3 tablespoons sugar
6–8 slices of white bread

1 Put the fruit, water and sugar in a small saucepan and bring to the boil. Reduce the heat, cover and simmer for a few minutes.
2 Cut the crusts off the bread and discard. Line the bottom and sides of a 1 kg (2 lb) pudding basin with most of the bread slices.
3 Fill with the fruit mixture and cover with the rest of the bread. Cover with a saucer, put a weight on top and chill in the refrigerator overnight.
4 Turn the pudding out on to a plate and serve with Greek yoghurt.
Note: It is important to use the correct size of basin. The pudding should fit into it so that it reaches the top. The saucer and weight prevent the pudding expanding as the bread soaks up the fruit juice, and makes sure the pudding is firm.

DAY 15

Breakfast
Toast with sugar-free jam

Mid-morning snack

Lunch
Avocado and Chive Soup*
Pepper and Lentil Salad*

Mid-afternoon snack

Dinner
Chicken and Parsley Rolls*
Courgettes
Carrots
Potatoes

Dessert
Orange Jelly*

Honey is a mixture of sugars, including mainly fructose and glucose. As fructose, weight for weight, tastes sweeter than sucrose, honey tastes that much sweeter. Nutritionally, however, there is no difference between sucrose and honey. One may, however, need to add less honey than sucrose to tea and other hot drinks to obtain the same degree of sweetness and, in this way, consume fewer calories.

Avocado and Chive Soup

serves 4

2 ripe avocados, stoned, peeled and roughly chopped
2 tablespoons lemon juice
900 ml (1½ pints) vegetable stock
1 tablespoon double cream
4 tablespoons snipped fresh chives

1 Purée the avocados in a blender or food processor with the lemon juice and vegetable stock. Transfer to a saucepan.
2 Add the cream and chives and mix thoroughly. Heat gently, without boiling, and serve hot.

Variation
This soup can also be served cold. Simply stir in the cream and chives and serve.

Pepper and Lentil Salad

serves 4

1 large red pepper, de-seeded and chopped
1 large green pepper, de-seeded and chopped
1 onion, finely chopped
10 pitted black olives, halved
1 tablespoon chopped fresh thyme
50 g (2 oz) Puy lentils, cooked
4 tablespoons olive oil
4 tablespoons lemon juice
1 garlic clove, crushed
black pepper, to taste

1 In a large bowl, mix together the red pepper, green pepper, onion, olives, thyme and lentils.
2 Whisk the oil, lemon, garlic and black pepper together and pour on to the lentil mixture. Mix well to coat all the ingredients.

Chicken and Parsley Rolls

serves 4

4 skinless chicken breast fillets, each weighing 100–150 g
(4–5 oz)
2 garlic cloves, crushed (optional)
4 tablespoons chopped fresh parsley

1 Put the chicken fillets on a firm surface, cover with damp greaseproof paper and beat with a meat mallet until evenly flattened.
2 Mix the garlic, if using, and parsley together and spread 1 tablespoon of the mixture on each chicken fillet. Roll up tightly and secure with cocktail sticks.
3 Place the chicken rolls in an ovenproof dish, cover and bake in a preheated oven at 180°C (350°F) mark 4 for 25 minutes or until the chicken is tender.
4 Remove the cocktail sticks, slice the chicken rolls and serve with the sauce of your choice.

Orange Jelly

serves 4

11 g (0.4 oz) sachet of powdered gelatine
600 ml (1 pint) unsweetened orange juice

1 Put 75 ml (3 fl oz) of the orange juice in a small heat-proof bowl and sprinkle over the gelatine. Leave to soften for 5 minutes, then stand the bowl in a saucepan of water and heat gently, stirring, until the gelatine is dissolved.
2 Add to the rest of the orange juice, pour into a wetted mould and chill in the refrigerator until set.

DAY 16

Breakfast

Corn Flakes, milk and chopped fresh fruit

Mid-morning snack

Lunch

Salad Niçoise*
Rice cakes

Mid-afternoon snack

Dinner

Beef Kebabs with Mint Yoghurt*
Brown rice

Dessert

Gooseberry Crumble* (see page 144)

Salad Niçoise

serves 4

4 medium potatoes, scrubbed
4 medium tomatoes, cut into wedges
200 g (7 oz) can tuna in oil, drained and flaked
1 small crisp lettuce, separated into leaves
black pepper, to taste
3 hard-boiled eggs, shelled and cut into wedges
50 g (2 oz) can anchovies in oil, drained
10 black olives
salad dressing (see page 118)

1 Cook the potatoes in boiling water for 15–20 minutes or until tender. Drain, cool, then cut into bite-sized pieces.
2 Mix the potato, tomatoes and tuna together.
3 Arrange the lettuce on a serving platter and spoon the

tuna mixture into the middle. Sprinkle with black pepper.
4 Arrange the egg, anchovies and olives on top of the salad and pour over a dressing of your choice.

Beef Kebabs with Mint Yoghurt

serves 4

450 g (1 lb) beef fillet or rump
2 tablespoons olive oil
2 tablespoons orange juice
1 garlic clove, crushed
black pepper, to taste
225 g (8 oz) button onions, skinned
2 courgettes
1 green pepper, de-seeded and cut into 2.5 cm (1 inch) pieces
1 red pepper, de-seeded and cut into 2.5 cm (1 inch) pieces
8 button mushrooms
2 teaspoons finely chopped fresh mint
225 g (8 oz) natural yoghurt

1 Cut the beef into 2.5 cm (1 inch) cubes. Mix the oil, orange juice, garlic and black pepper in a shallow dish, add the meat and leave to marinate for 2 hours.
2 Blanch the onions in boiling water for 3 minutes, then drain. Cut the courgettes into 2.5 cm (1 inch) pieces.
3 Drain the beef, reserving the marinade. Thread the beef and vegetables on to eight kebab skewers.
4 Cook under a preheated grill for 10–15 minutes, basting occasionally with the marinade.
5 While the kebabs are cooking, prepare the yoghurt dressing by mixing the yoghurt and mint together.
6 When the kebabs are cooked, serve with the dressing.

The same weight of sugar and protein contain similar amounts of energy; both contain approximately half the energy of fats.

DAY 17

Breakfast
Yoghurt, chopped banana and 1 tablespoon seeds

Mid-morning snack

Lunch
Cheese salad sandwich

Mid-afternoon snack

Dinner
Tofu, Bean and Herb Stir-Fry*
Tropical Rice Salad* (see page 112)

Dessert
Ginger Fruit Salad*

Tofu, Bean and Herb Stir-Fry
serves 4

2 tablespoons vegetable oil
275 g (10 oz) tofu (bean curd), drained, dried and cut into
cubes
2 garlic cloves, crushed
350 g (12 oz) green beans
3 tablespoons chopped fresh herbs (such as thyme, parsley,
chervil and chives)
4 spring onions, thinly sliced
2 tablespoons soy sauce

1 Heat 1 tablespoon of the oil in a wok or frying pan.
2 When the oil is hot, add the tofu and garlic and stir-fry
for 2 minutes. Lift out with a slotted spoon and drain.

3 Heat the remaining oil in the pan and, when hot, add the green beans and stir-fry gently for 4–5 minutes.
4 Add the herbs, spring onions and soy sauce, and stir-fry for a further minute.
5 Return the tofu to the pan and heat through for 1 minute, then serve immediately.

Ginger Fruit Salad

serves 4–6

juice of 2 oranges
3 tablespoons juice from stem ginger jar
25 g (1 oz) light soft brown sugar
2 oranges
1 ogen melon, weighing about 1 kg (2 lb)
3 dessert apples
2 pieces of preserved stem ginger, finely chopped
fresh mint leaves, to decorate

1 Put the orange juice in a saucepan with the stem ginger juice and sugar. Heat gently until the sugar has dissolved. Bring to the boil and boil for about 5 minutes or until syrupy. Remove from the heat and leave to cool.
2 Peel the oranges, removing the pith, and slice into thick rings. Cut the rings into quarters, removing any pips.
3 Cut the melon in half, remove the seeds and scoop out the flesh, using a melon baller if possible.
4 Peel and core the apples, then chop roughly.
5 Put all the fruit into a serving bowl, or into the empty melon halves, with the ginger and sugar syrup. Stir well, cover and chill for 1–2 hours. Decorate with mint.

In the mouth, the sticky sugars that adhere to the teeth are broken down into acids which dissolve protective dental enamel and produce dental caries.

DAY 18

Breakfast
Scrambled egg on toast

Mid-morning snack

Lunch
Pepper and Lentil Salad* (see page 150)
Apple and Nut Salad* (see page 124)

Mid-afternoon snack

Dinner
Yoghurt Roast Chicken*
Potatoes
Mange-tout
Broccoli
Sweetcorn

Dessert
Dried Fruit Compote*

Yoghurt Roast Chicken

serves 4

225 ml (8 fl oz) natural yoghurt
2 teaspoons curry powder (mild, medium or hot, as
preferred)
½ bunch fresh coriander, finely chopped
1 tablespoon grated fresh root ginger
3 garlic cloves, crushed
ovenready chicken, weighing 1–1.5 kg (3–4 lb)

1 Combine the yoghurt, curry powder, coriander, ginger
and garlic. Rub all over the chicken, cover and set aside
for 1–2 hours.

2 Place the chicken in a roasting tin with the yoghurt sauce and bake in a preheated oven at 180°C (350°F) mark 4 for 1¼–1½ hours.

3 Transfer the chicken to a serving dish and keep warm. Stir together the chicken juices and yoghurt sauce remaining in the roasting tin and boil rapidly to reduce, if necessary. Serve the chicken with the sauce.

Dried Fruit Compote

serves 2

100 g (4 oz) mixed dried fruit, such as peaches, prunes,
apples, apricots and pears
100 ml (4 fl oz) orange juice
2 whole cloves
5 cm (2 inch) cinnamon stick
grated rind and juice of ½ lemon

1 Wash the fruit and put it in a bowl with the orange juice, spices, lemon rind and juice. Leave to soak overnight.

2 Next day, if the juice has been absorbed, add 2 tablespoons water, then place the mixture in a saucepan. Bring to the boil, then reduce the heat, cover and simmer for 10–15 minutes.

3 Transfer the fruit to a serving bowl and remove the cinnamon and cloves. Serve warm or leave to cool.

Sugar itself is neither acid nor alkaline but neutral, like water. However, once acted upon by bacteria a variety of acids are produced. Fermentation of sugar by yeast in the presence of oxygen from the air does not result in alcohol being produced but acid, hence wine turning to vinegar.

DAY 19

Breakfast
Oats (soaked overnight) with milk, chopped fresh fruit
and 1 tablespoon chopped nuts

Mid-morning snack

Lunch
150-175 g (5–6 oz) jacket potato
Coleslaw Salad* (see page 142)

Mid-afternoon snack

Dinner
Quick Prawn Curry*
Brown rice
French beans

Dessert
Summer Pudding* (see page 148)

Quick Prawn Curry

serves 4

1 tablespoon sunflower oil
1 medium onion, chopped
2 garlic cloves, crushed
1 tablespoon plain flour
2 teaspoons curry powder
475 ml (16 fl oz) water
1 tablespoon tomato purée
1 medium potato, peeled and sliced
275 g (10 oz) peeled cooked prawns

1 Heat the oil in a frying pan and gently fry the onion and garlic for 2–3 minutes.

2 Add the flour and curry powder, mix well, and cook, stirring, for 1 minute.

3 Add the water and tomato purée and continue to cook, stirring, until the mixture thickens. Add the potato, cover and simmer for 15 minutes.

4 Add the prawns and simmer for a further 5 minutes or until the potato is tender. Serve with brown rice or spinach pasta.

Yeast bacteria, as well as all kinds of other germs, have nutrient requirements just as human beings do. Although yeast can live off sugar, it cannot do so if the sugar concentration is too high. Too much sugar is also lethal for humans and inhibits the growth of many bacteria, hence the use of sugar to preserve fruits, as in jams, marmalade, and canned and bottled fruits.

DAY 20

Breakfast
Grilled tomatoes on toast
Grilled mushrooms

Mid-morning snack

Lunch
Sardines in oil, drained
Salad of choice
Potato Salad* (see page 119)

Mid-afternoon snack

Dinner
Peach and Ham-Stuffed Drumsticks*
Leek Gratin*
Sauté potatoes

Dessert
Blackcurrant Sorbet* (see page 116)

Peach and Ham-stuffed Drumsticks

serves 3

6 large chicken drumsticks, boned
6 slices of lean ham
3 canned peach halves, drained and halved
black pepper, to taste

1 Place a slice of ham and a piece of peach on each piece of chicken and sprinkle with black pepper. Roll up and place in a baking dish.
2 Bake in a prehated oven at 180°C (350°F) mark 4 for 45 minutes or until tender.

Leek Gratin

serves 4–6

1 tablespoon sunflower oil
1 onion, finely chopped
2 garlic cloves, crushed
275 g (10 oz) button mushrooms
4 small courgettes, roughly chopped
100 g (4 oz) fresh white breadcrumbs
1 tablespoon tomato purée
black pepper, to taste
3 large leeks, sliced
100 g (4 oz) French beans
1 quantity Light Cheese Sauce (see page 147)
75 g (3 oz) grated low-fat Cheddar cheese

1 Heat the oil in a frying pan and gently fry the onion and garlic for 2–3 minutes.
2 Add the mushrooms and courgettes and cook for a further 2 minutes. Stir in the breadcrumbs, tomato purée and black pepper, then remove from the heat and cover.
3 Cook the leeks and French beans in boiling water for 15 minutes or until just tender. Drain, place in an ovenproof dish and cover with the mushroom mixture.
4 Heat the cheese sauce gently and pour it over the leek and mushroom mixture. Sprinkle with the cheese and bake in a preheated oven at 200°C (400°F) mark 6 for 20 minutes.

It is not just sugar that tastes sweet. Taste buds on the tongue have a particular shape and structure to them that allows them to respond to the sugar molecule. The same stimulus may be achieved with artificial sweeteners and even some mineral preparations.

DAY 21

Breakfast
Grilled bacon, tomatoes and mushrooms
Toast

Mid-morning snack

Lunch
Jacket potato, baked beans and salad

Mid-afternoon snack

Dinner
Cold roast lamb with Yoghurt and Cucumber Dip*
Tropical Rice Salad* (see page 112)
Coleslaw Salad (see page 142)

Dessert
Pineapple Water-Ice (see page 126)*

Yoghurt and Cucumber Dip

serves 4

½ medium cucumber, skinned and roughly chopped
150 ml (5 fl oz) Greek yoghurt
1 garlic clove, crushed
1 tablespoon chopped fresh mint

Mix all the ingredients together and chill before serving.

Chocolate contains several active chemicals, including caffeine and a further mood-stimulating substance, beta phynyl ethylamine.

DAY 22

Breakfast

1 poached egg on toast
2 rice cakes and sugar-free marmalade

Mid-morning snack

Lunch

Salmon Steaks with Ginger*
Endive, Fruit and Nut Salad* (see page 128)
Potato Salad* (see page 119)

Mid-afternoon snack

Dinner

Pork with Peppers and Black Bean Sauce*
Brown rice

Dessert

Canned pineapple (in natural juice) with yoghurt

Salmon Steaks with Ginger

serves 2

2 salmon steaks
2 tablespoons lemon juice
2.5 cm (1 inch) square of fresh root ginger, peeled and
finely chopped
black pepper, to taste

1 Place each salmon steak on a large piece of foil. Add 1
tablespoon lemon juice and half the chopped ginger to
each steak. Season with a little black pepper.
2 Wrap the steaks individually in foil to make two parcels
and bake in a preheated oven at 180°C (350°F) mark 4 for
20 minutes. Serve hot with vegetables or cold with salad.

Pork with Peppers and Black Bean Sauce

serves 4

3 tablespoons black beans
1 tablespoon vegetable oil
1 green pepper, de-seeded and sliced
1 red pepper, de-seeded and sliced
1 tablespoon dry sherry
2 tablepoons light soy sauce
2 garlic cloves, crushed
1 tablespoon cornflour
175 ml (6 fl oz) water
225 g (8 oz) lean pork, cut into 2.5 cm (1 inch) cubes

1 Rinse the black beans in cold water, dry them, then crush them in a pestle and mortar or with a spoon.
2 Heat the oil in a wok or frying pan and gently fry the green and red peppers for 2–3 minutes. Remove with a slotted spoon and set aside.
3 Mix together the black beans, sherry, soy sauce and garlic. Mix the cornflour with 2 tablespoons water and stir into the black bean mixture with the rest of the water.
4 Put the pork in the wok or frying pan and stir-fry until evenly browned. Add the sauce and bring to the boil. Reduce the heat, cover and simmer for about 30 minutes or until the pork is tender, stirring occasionally. Do not reduce too much; add extra water if necessary.
5 Before serving, stir in the peppers and heat through.

Molasses from cane sugar contains some important trace minerals, including chromium which helps with the control of blood sugar levels and metabolism.

DAY 23

Breakfast
Rice Krispies with milk and 1 tablespoon seeds

Mid-morning snack

Lunch
Watercress Soup*
Jacket potato
Coleslaw Salad* (see page 142)

Mid-afternoon snack

Dinner
Creamy Chicken Curry*
Brown rice
Broccoli

Dessert
Frozen Peach Treat*

When the highly poisonous mineral, lead, is combined with an acid, like vinegar, it produces 'sugar of lead'. This sweet-tasting compound was formed and indeed even consumed in Roman times when acidic wine was put into lead containers to 'sweeten' it. Such a practice made adding sugar to wine a much more attractive proposition.

Watercress Soup

serves 4

50 g (2 oz) vegetable margarine
1 medium onion, finely chopped
225 g (8 oz) potatoes, peeled and diced
600 ml (1 pint) chicken stock
pinch of grated nutmeg
4 bunches of watercress, washed and trimmed
black pepper, to taste
2 tablespoons single cream

1 Melt the margarine in a frying pan, add the onion and fry gently for about 5 minutes or until the onion is transparent.
2 Add the potato, chicken stock and nutmeg and simmer for 15 minutes.
3 Add the watercress and simmer for a further 10 minutes. Purée the soup in a blender or food processor, then push through a sieve. Add black pepper to taste and the cream. Serve hot or chilled.

Creamy Chicken Curry

serves 4

1 ovenready chicken, weighing about 1.5 kg (3 lb)
1 tablespoon vegetable oil
1 onion, finely chopped
2 garlic cloves, crushed
½ teaspoon grated fresh root ginger
2 tablespoons curry powder (mild, medium or hot, as preferred)
1 tablespoon vinegar
150 ml (¼ pint) milk
150 ml (¼ pint) coconut cream
1 tablespoon sesame seeds

1 Cut the chicken into four pieces and remove the skin.
2 Heat the oil in a flameproof casserole, add the onion, garlic and ginger and cook gently for 8–10 minutes, stirring continuously. Add the curry powder and vinegar and cook over a high heat for 1 minute.
3 Add the chicken, milk and coconut cream and mix well. Simmer gently for 20–30 minutes or until the chicken is tender. Serve sprinkled with sesame seeds.

Frozen Peach Treat

serves 4

3 ripe peaches, skinned, stoned and chopped
600 ml (1 pint) natural yoghurt
4 teaspoons lemon juice
honey, to taste

1 In a large bowl, mash the peaches with a potato masher. (If you use a blender, more of the water will be drawn out from the fruit and you will have more crunchy water crystals.)
2 Add the yoghurt, lemon juice and honey and stir.
3 Spoon into freezer-proof bowls or cups and freeze for 2–3 hours.

Contrary to popular belief, there is often little difference between brown sugar and white sugar. This is certainly true with regard to effects on the metabolism and dental caries. Brown sugar, however, is less refined; it has been through fewer manufacturing processes than white.

DAY 24

Breakfast

Yoghurt, chopped fresh fruit and seeds

Mid-morning snack

Lunch

Turkey and coleslaw sandwich

Mid-afternoon snack

Dinner

Tandoori Mackerel*

Sauté potatoes, sweetcorn and spinach

Dessert

Baked Apple* (see page 120)

Tandoori Mackerel

serves 2

2 mackerel, cleaned

3 tablespoons tandoori paste

300 ml (½ pint) natural yoghurt

1 tablespoon lemon juice

1 Make four slashes in both sides of each fish.
2 Mix the tandoori paste, yoghurt and lemon juice.
3 Place the mackerel in a shallow dish and pour over the tandoori mixture. Turn the fish so that all sides are well coated. Leave to marinate for at least 1–2 hours.
4 Grill the fish for 5 minutes on each side, or until tender.

There are approximately 7 teaspoons of sugar in a can of lemonade and 8 teaspoons in a can of cola.

DAY 25

Breakfast
Grilled lean bacon sandwich

Mid-morning snack

Lunch
Jacket potato
Prawns
Salad of choice

Mid-afternoon snack

Dinner
Grilled gammon
Mushrooms with Buckwheat Stuffing*
Boiled potatoes
Fresh greens

Dessert
Dried Fruit Compote* (see page 157)

It is possible to become allergic to sugar, or to develop an intolerance. Some people are unable to digest sucrose and this can lead to diarrhoea; others are unable to digest fructose (fruit sugar) as well as sucrose and this can cause diarrhoea and other abdominal symptoms. These effects are experienced only by one in several thousand of the normal population, but they may run in families.

Mushrooms with Buckwheat Stuffing

serves 4

2 teaspoons vegetable oil
1 medium onion, finely chopped
600 ml (1 pint) vegetable stock
200 g (7 oz) buckwheat
75 g (3 oz) cooked brown rice
1 tablespoon chopped fresh herbs, such as sage and parsley
black pepper, to taste
4 large or 8 medium open-cup mushrooms, stalks removed
50 g (2 oz) low-fat cheese, finely grated

1 Heat the oil in a frying pan and gently fry the onion for about 5 minutes or until transparent.
2 Add the vegetable stock and bring to the boil, then remove from the heat.
3 Add the buckwheat, brown rice and herbs to the stock. Season with pepper and mix well.
4 Lay the mushrooms in a baking tray and fill them with the buckwheat mixture. Sprinkle over the cheese and bake in a preheated oven at 180°C (350°F) mark 4 for 15 minutes.

Variations
This mixture can also be used to stuff cabbage leaves, peppers, courgettes or tomatoes.

DAY 26

Breakfast
Boiled egg
Toast

Mid-morning snack

Lunch
Tuna salad sandwich

Mid-afternoon snack

Dinner
Mustard Cod Crêpes*
Spinach
Mixed Root Vegetables*

Dessert
Grape and White Wine Jelly*

High sugar consumers are known to exhibit a number of subtle changes in metabolism. These include increased demand for vitamin B_1 (thiamin), increased losses of the mineral chromium in the urine, and it may affect the balance of other minerals. These effects occur silently and their significance is dependent upon the quality of the individual's diet as well as their general level of health.

Mustard Cod Crêpes

makes 12 crêpes

For the crêpes
75 g (3 oz) plain wholemeal flour
75 g (3 oz) plain white flour
1 large egg (size 1), beaten
450 ml (¾ pint) skimmed milk
2-3 tablespoons sunflower oil

For the filling
350 g (12 oz) cod fillet
50 g (2 oz) vegetable margarine
25 g (1 oz) plain flour
450 ml (¾ pint) milk
juice of 1 lemon
black pepper, to taste
2 tablespoons mustard
lemon slices, to garnish

1 To make the crêpes, put the flours in a bowl and make a well in the centre.
2 Add the egg and half the milk and gradually mix in the flour from the sides of the bowl, whisking until smooth. Slowly whisk in the remaining milk. Add a little extra milk if the batter seems thick.
3 Heat 1 teaspoon oil in a small non-stick frying pan and pour in 3 tablespoons of the batter. Tilt the pan so that the batter spreads evenly over the base, then cook over a medium heat for 2 minutes or until golden brown underneath.
4 Toss or turn the crêpe and cook for a further 2 minutes. Slide the crêpe on to a plate, cover with greaseproof paper and wrap in a clean cloth.
5 Continue until you have made 12 crêpes. Pile them one on top of the other with greaseproof paper in between them, and keep wrapped in a clean cloth until required.
6 To make the filling, steam the cod over a pan of boiling

water on a covered plate with 1 tablespoon milk, or in a steamer for 10–15 minutes or until the fish flakes easily.

7 Melt the margarine in a saucepan, stir in the flour and cook for 1 minute. Remove from the heat and gradually stir in the milk. Return to the heat and slowly bring to the boil, stirring constantly. Simmer slowly for about 3 minutes, stirring.

8 Stir the lemon juice into the sauce and season with black pepper. Mix well, stir in the mustard and remove from the heat. Spoon half the sauce into a bowl.

9 Remove any skin from the cod and flake into small pieces. Add to half the sauce, reserving the other half.

10 Fill each crêpe with 2 tablespoons cod mixture and roll up.

11 Carefully place the crepes in an ovenproof dish and cover with the remaining sauce. Bake in a preheated oven at 170°C (325°F) mark 3 for 10 minutes or until heated through. Serve immediately, garnished with lemon slices.

Mixed Root Vegetables

serves 4

3 medium turnips, peeled and cut into chunks
4 large carrots, roughly sliced
½ medium swede, peeled and cut into chunks
15 g (½ oz) vegetable margarine
black pepper, to taste

1 Steam or boil all the vegetables until soft, then drain, if necessary.

2 Mash the vegetables together with the margarine and black pepper, to taste. Alternatively, purée the vegetables in a blender or food processor with the margarine and pepper. Serve at once.

Grape and White Wine Jelly

serves 4

300 ml (½ pint) diluted apple juice
4 teaspoons powdered gelatine
300 ml (½ pint) dry white wine
225 g (8 oz) green grapes, peeled, halved and de-seeded
(if necessary)
green grapes, to decorate

1 Put 3 tablespoons of the apple juice in a small heatproof bowl and sprinkle in the gelatine. Leave to soften for 5 minutes, then stand the bowl in a saucepan of hot water and heat gently, stirring, until the gelatine has dissolved.
2 Combine the remaining apple juice and the white wine with the gelatine mixture and leave until syrupy.
3 Stir the prepared grapes into the jelly and pour into a 900 ml (1½ pint) jelly mould or four individual moulds. Chill in the refrigerator for 2–3 hours or until set.
4 Carefully turn the jelly out on to a serving plate and decorate with grapes.

DAY 27

Breakfast

Oats (soaked overnight) with mixed chopped fresh fruit

Mid-morning snack

Lunch

Quick Ham Risotto*

Mid-afternoon snack

Dinner

Mushroom and Spinach Yoghurt Quiche*
Apple and Nut Salad* (see page 124)
Potato Salad* (see page 119)

Dessert

Greek yoghurt with chopped nuts and 1 teaspoon honey

Quick Ham Risotto

serves 4

4 leeks, chopped
1 tablespoon sunflower oil
100 g (4 oz) brown long-grain rice, cooked
100 g (4 oz) frozen peas
100 g (4 oz) frozen sweetcorn
1 red pepper, de-seeded and chopped
1 teaspoon grated orange rind
½ teaspoon cayenne pepper
½ teaspoon ground cumin
black pepper, to taste
450 g (1 lb) lean cooked ham, cubed

1 Steam the leeks for about 3 minutes.
2 Heat the oil in a frying pan and add the cooked rice, peas, sweetcorn, red pepper, orange rind, cayenne pepper, cumin, leeks and black pepper. Cook for 2–3 minutes.
3 Add the ham and heat through for 2–3 minutes.

Mushroom and Spinach Yoghurt Quiche

serves 4–6

225 g (8 oz) shortcrust pastry, thawed if frozen
1 tablespoon vegetable oil
100 g (4 oz) mushrooms, sliced
1 medium onion, sliced
2 eggs, beaten
150 g (5 oz) natural yoghurt
150 ml (¼ pint) milk
225 g (8 oz) frozen chopped spinach, thawed and drained
black pepper, to taste

1 Roll out the pastry and use to line a 24 cm (9½ inch) loose-bottomed flan ring. Bake blind in a preheated oven at 200°C (400°F) mark 6 for 10–15 minutes.
2 Heat the oil in a frying pan and lightly fry the mushrooms and onion for 2–3 minutes.
3 Mix the eggs with the yogurt, milk, spinach and black pepper, then add the mushroom and onion mixture.
4 Pour into the pastry case and bake in the oven at 190°C (375°F) mark 5 for 30 minutes or until just set.

Increased consumption of sucrose has been partly blamed for conditions such as obesity, diabetes, gall stones, peptic ulcers and hypoglycaemia.

DAY 28

Breakfast
Scrambled eggs
Grilled tomatoes
Rice cakes

Mid-morning snack

Lunch
Jacket potato
Cottage cheese with chopped pineapple
Green salad

Mid-afternoon snack

Dinner
Grilled chicken
Plum and Ginger Sauce*
Cabbage
Carrots
Brown rice

Dessert
Jellied Grapefruit*

Plum and Ginger Sauce

serves 4

2 teaspoons sunflower oil
1 small onion, finely chopped
1 garlic clove, crushed
1 teaspoon grated fresh root ginger
3 tablespoons water
2 canned plums (in natural juice)
3 tablespoons plum juice from can

1 Heat the oil in a saucepan, add the onion, garlic and ginger and fry for 2–3 minutes.
2 Add the remaining ingredients and bring to the boil, then remove from the heat and leave to cool slightly.
3 Purée the sauce in a blender or food processor, then re-heat before serving, if required.

Jellied Grapefruit

serves 4

2 large pink grapefruit, halved
300 ml (½ pint) unsweetened grapefruit juice
300 ml (½ pint) unsweetened orange juice
11 g (0.4 oz) sachet powdered gelatine

1 Scoop the grapefruit segments out of the grapefruit halves and remove any pith from the segments and from the grapefruit shells. Reserve any juice.
2 Place the four grapefruit shells in a dish and divide the grapefruit segments evenly between them.
3 Mix the reserved fresh and the unsweetened juices and put 2 tablespoons in a small heatproof bowl.
4 Sprinkle the gelatine into the fruit juice mixture and leave to soften for 5 minutes. Stand the bowl in a saucepan of hot water and heat gently, stirring, until the gelatine has dissolved.
5 Mix the gelatine with the remaining fruit juice mixture and pour equal amounts into each grapefruit shell. Place in the refrigerator to set for 1–2 hours, before serving.

A number of myths have grown up relating to the adverse health aspects of sucrose. There is little evidence that moderate sucrose consumption by itself is the cause of heart disease, high blood pressure, acne or even diabetes that requires insulin.

DAY 29

Breakfast

Mixed fruit with 1 tablespoon linseeds and 1 tablespoon chopped nuts
Rice cakes with sugar-free jam

Mid-morning snack

Lunch

Jacket potato
Steamed mixed vegetables
Light Cheese Sauce* (see page 147)

Mid-afternoon snack

Dinner

Seafood Salad* (see page 136)
Pasta Salad* (see page 136)

Dessert

Gooseberry Mousse*

Gooseberry Mousse

serves 4

450 g (1 lb) gooseberries, topped and tailed
concentrated apple juice, to taste
1 tablespoon lime juice
300 ml (½ pint) unsweetened apple purée
1 tablespoon clear honey
3 tablespoons natural yoghurt
2 tablespoons water
3 teaspoons powdered gelatine
1 egg white

1 Put the gooseberries in a saucepan with a little water and simmer gently until lightly stewed. Add concentrated apple juice to taste.

2 Put the gooseberries and lime juice in a blender or food processor and purée until smooth. Pour the purée into a measuring jug and make it up to 300 ml (½ pint) by adding water.

3 Mix the gooseberry and the apple purées together and add the honey and yoghurt.

4 Put the water in a small heatproof bowl and sprinkle in the gelatine. Leave to soften for 5 minutes, then stand the bowl in a saucepan of hot water and heat gently, stirring, until the gelatine dissolved. Stir the gelatine into the fruit purée.

5 Whisk the egg white until stiff and fold lightly into the gooseberry and apple mixture. Spoon into four dessert glasses and chill for at least 2–3 hours before serving.

Chromium is a mineral which is thought to play a very important part in the regulation of blood glucose in the body. Only minute quantities are required by us and a lifetime's supply of chromium weighs only a fraction of an ounce.

DAY 30

Breakfast

Yoghurt with chopped fresh fruit and 2 tablespoons
linseeds

Mid-morning snack

Lunch

Ryvita
Yoghurt and Cucumber Dip* (see page 162)
Endive, Fruit and Nut Salad* (see page 128)

Mid-afternoon snack

Dinner

Grilled mackerel
Butter beans
Spring greens
Sauté Potatoes

Dessert

Stewed fruit with Frumble Topping* (see page 138)

Babies are born with a good supply of the mineral
chromium, but levels of chromium in the body are
depleted as we get older. A diet high in refined
processed foods can cause chromium to be lost
through the urine.

Recommended Reading

Recipe Books

The Allergy Cook Book, Danila Armstrong, SRD, and Dr Andrew Cant, Octopus Books

The Make It Simple Cook Book, Weight Watchers' Step-by-Step Guide to Easy Cooking, Anne Page-Wood, New English Library

The Slimmer's Year, Anne Ager, Julie Hamilton and Miriam Polunin, Hamlyn

Seasonal Salads, David Scott and Paddy Byrne, Ebury Press

Calorie-Counted Meals, Alex Barber, J Sainsbury

Raw Energy Recipes, Leslie and Susannah Kenton, Century

Beat PMT Cookbook, Maryon Stewart and Sarah Tooley, Ebury Press

The Sainsbury Book of Salads, Carole Handslip, J Sainsbury

The Reluctant Vegetarian, Simon Hope, William Heinemann

Gourmet Vegetarian Cooking, Rose Elliot, Fontana

Healthy Cooking, Tesco Stores

The Single Vegan, Leah Leneman, Thorsons

Bumper Bake Book, Rita Greer, Bunterbird Ltd

General

Escape from Tranquillisers and Sleeping Pills, Larry Neild, Ebury Press

Do-it-yourself Shiatsu, W. Ohashi, Unwin

Shopping for Health, Jannette Marshall, Pan Books

Alternative Health Care for Women, Patsy Westcott, Thorsons

The Well Woman's Self-Help Book, Nikki Bradford, Sidgewick and Jackson

The Book of Massage, Clare Maxwell-Hudson, Ebury Press

Food Irradiation: The Facts, Tony Webb and Dr Tim Lang, Thorsons

Parents For Safe Food Handbook, edited by Joan and Derek Taylor, Ebury Press

Nutritional Medicine, Dr Stephen Davies and Dr Alan Stewart, Pan Books

Beat PMT Through Diet, Maryon Stewart with contributions from Dr Guy Abraham and Dr Alan Stewart, Ebury Press

Candida – Diet Against It, Luc de Schepper, Foulsham

Quantum Carrot, A New Concept in Small Space Organic Gardening, Branton Kenton, Ebury Press

The Food Scandal, Caroline Walker and Geoffrey Cannon, Century

Pure, White and Deadly, Professor John Yudkin, Viking

Candida Albicans: Could Yeast be Your Problem, Leon Chaitow, Thorsons

Diet Books

The Vitality Plan, Maryon Stewart and Dr Alan Stewart, Little Brown Books (December 1992)

The Allergy Diet, Elizabeth Workman, SRD, Dr John Hunter and Dr Virginia Alun Jones, Martin Dunitz

Food Combining For Health, Doris Grant and Jean Joice, Thorsons

Raw Energy, Leslie and Susannah Kenton, Century

The Real Food Shop and Restaurant Guide, Clive Johnstone, Ebury Press

Organic Consumer Guide/Food You Can Trust, edited by David Mabey and Alan and Jackie Gear, Thorsons

Stress

Self Help for Your Nerves, Dr Clair Weekes, Angus and Robertson

Mind Power, Dr Vernon Coleman, Vermilion

Specialised Food and Nutritional Supplement Suppliers

UK

Boots the Chemist
  (large stores)

Efamol and Optivite

Foodwatch International
9 Corporation Street
Taunton
Somerset TA1 4AJ
Tel: 0823 325022

Sugar-free jelly, ice cream mix and just about everything.

Mail order service

Granose Foods
Stanborough Park
Watford
Herts WD2 6JR
Tel: 0923 672281

Produce organically grown baby foods with no added sugar or salt. Other products gluten or lactose free (or both). Special products for adults.

Nature's Best Ltd
PO Box 1
Tunbridge Wells
Kent TN2 3EQ
Tel: 0892 34143

Mail order nutritional supplements.

Nutritional Health Ltd.
PO Box 713
Hove
East Sussex BN3 5DT
Tel: 0273 771366

Specialised flour mixes for those avoiding wheat – to make cakes, muffins, pizza, pasta, pastry and biscuits. Also suppliers of specialised books and nutritional supplements, including Optivite, Sugar Factor and Health Insurance Plus

AUSTRALIA

NNFA (National Nutrition
   Foods Association)
PO Box 84
Westmead
NSW 2145
Tel: (02) 633 9913

The NNFA has lists of all supplement stocklists and retailers in Australia if you have any difficulty obtaining supplements.

Suppliers of Sugar Factor.

Blackmores
23 Roseberry Street
Balgowlah
NSW 2093
Tel: (02) 949 3177

NEW ZEALAND

NNFA (National Nutrition
   Foods Association)
PO Box 820062
Auckland

The NNFA has lists of all supplement stocklists and retailers in New Zealand if you have any difficulty obtaining supplements.

Suppliers of Sugar Factor.

Blackmores
2 Parkhead Place
Albany
Auckland
Tel: (09) 415 8585

Useful Addresses

Women's Health

Women's Nutritional Advisory Service, PO Box 268, Hove, East Sussex BN3 1RW
Tel: 0273 771366
(Please send a stamped addressed envelope.)

Anorexic Support, Norah Robinson-Smith, 15 Sharnford Road, Leicester
Tel: 045-527 2398

Brook Advisory Centres, 223 Tottenham Court Road, London W1P 9AE
Tel: 071-580 2911/323 1522

Natural Family Planning Service, Catholic Marriage Advisory Council, 15 Lansdowne Road, London W11 3AJ
Tel: 071-727 0141

Women's Health Information Centre, 52 Featherstone Street, London EC1
Tel: 071-251 6580

Alternative Health

Migraine Trust, 45 Great Ormond Street, London WC1N 3HD
Tel: 071-278 2676

British Society for Nutritional Medicine, PO Box 3AP, London W1A 3AP
Tel: 071-436 8532
(Your GP will have to contact BSNM for you.)

British Homeopathic Association, 27a Devonshire Street, London W1N 1RJ
Tel: 071-935 2163

British Acupuncture Register and Directory, 34 Alderney Street, London SW1V 4UE
Tel: 071-834 1012

The Shiatsu Society, 14 Oakdene Road, Redhill, Surrey
Tel: 0737 767896

The Henry Doubleday Research Association, Ryton Gardens, National Centre for Organic Gardening, Ryton on Dunsmore, Coventry CV8 3LG
Tel: 0203-303517

The Council for Acupuncture, Suite 1, 19a Cavendish Square, London W1M 9AD
Tel: 071-409 1440

The European School of Osteopathy, 104 Tonbridge Road, Maidstone, Kent ME16 8SL
Tel: 0622-671558

The British School of Osteopathy, Little John House, 1-4 Suffolk Street, London SW1 4HG
Tel: 071-930 9254/8

General

Tranx (UK) Ltd, National Tranquilliser Advice Centre, Registered Office, 25a Masons Avenue, Harrow, Middlesex HA3 5AH
Tel: Client Line 081-427 2065
24-hour answering line: 081-427 2827

ASH (Action on Smoking and Health), 5-11 Mortimer Street, London W1N 7RH
Tel: 071-637 9843

Alcoholics Anonymous (AA), General Services Office, PO Box 1, Stonebow House, Stonebow, York YO1 2NJ
Tel: 0904-644026

Anorexia and Bulimia Nervosa Association, Tottenham Women's Health Centre, Annexe C, Tottenham Town Hall, Town Hall Approach, London N15 4RX
Tel: 081-885 3936 (Wednesdays 6-9pm only)

The Food Magazine, The Food Commission, 88 Old Street, London EC1
Tel: 071-253 9513

Friends of the Earth, 26 Underwood Street, London N1
Tel: 071-490 1555

Parents for Safe Food, Britannia House, 1/11 Glenthorne Road, London W6

Soil Association, 86/88 Colston Street, Bristol

Hyperactive Children Support Group, 59 Meadowside, Angmering, Littlehampton, West Sussex BN16 4BW
(postal enquiries only)

Relate, Herbert Gray College, Little Church Street, Rugby CV21 3AP
Tel: 0788-573241

National Childbirth Trust, Alexandra House, Oldham Terrace, London W3 6NH
Tel: 081-992 8637

Samaritans, 17 Uxbridge Road, Slough SL1 1SN
Tel: 0753-327133

Help for Health, The Grant Building, Southampton General
Hospital, Southampton SO9 4XY
(They have a large computer database of countrywide self-help
and support groups for almost every health problem imaginable.
Can provide names, addresses and telephone numbers of organisations.)

Weight Watchers UK Ltd., Kidwells Park House, Kidwells Park
Drive, Maidenhead, Berks SL6 8YT
Tel: 0628-777077

British Diabetic Association, 10 Queen Anne Street, London W1
Tel: 071-323 1531

British Wheel of Yoga, 1 Hamilton Place, Boston Road,
Sleaford, Lincs
Tel: 0529-306851

National Society for Non-Smokers, Latimer House, 40-48
Hanson Street, London W1P 7DE

Relaxation for Living, 29 Burwood Park Road, Walton-on-
Thames, Surrey KY12 5LH
(Please send a stamped addressed envelope.)

The Eating Disorders Association, Sackville Road, 44-46
Magdalen Street, Norwich, Norfolk NR3 1JE
(Please send a stamped addressed envelope.)

Parents Anonymous Lifeline, 6-9 Manor Gardens, London N7
Tel: 071-263 8918
(For parents having problems with children or babies of any
age. Please send a stamped addressed envelope.)

British Dental Health Foundation (BDHF), 88 Gunards Avenue,
Fishermead, Milton Keynes MK6 2BL
Tel: 0809-667063

Health Education Authority, Hamilton House, Mabledon Place,
London WC1H 9TX
Tel: 071-383 3833

Action & Information on Sugars, PO Box 459, London SE5
7QA

Women's Environmental Network, 287 City Road, London EC1

Index

acesulfame potassium 68, 69–70
adrenalin (epinephirne) 51
advertising
  Advertising Standards Authority
    (ASA) 31
  Cadbury 28
  campaign 27–8
  cinema 29
  cost of 28
  Independent Broadcasting Authority
    (IBA) 31
  Mars 29–30
  omissions in 30–1
  power of 28–30
  television 29, 31
alcohol 54, 97
altered glucose syrups 70
anorexia nervosa 80, 82–3
aspartame 68, 69

beet see sugar
blood sugar
  control of 50–1
  factors effecting 54
  glycaemic index 55
  hormones 50–1
  influence of food groups on 55–6
  influence of nutrients on 56–76
  insulin 50, 51, 53, 54, 56
  'Kreb's Cycle, The' 56
  levels 18, 19, 49–58, 86
  liver function 50
  low see hypoglycaemia
  WNAS Sugar Factor 58, 88
bulimia nervosa 80, 82–3

cancer 21, 70
cane see sugar
carbohydrates 17, 18, 25, 28, 50
cardiovascular disease 23, 24
cereal bars 61–3
'chocoholics' 38–42, 82
chocolate
  consumption 11–12, 38–43, 84
chromium 25, 57–8, 90
Chromium Glucose Tolerance Factor
  (GTF) 57
coffee drinking 54, 177, 85, 95
COMA report, Dietary Sugars and
  Human Disease 21, 22
corticosteroids (steroids) 51

dental caries see tooth decay
depression 33, 76

diabetes 23, 24–5, 50–1
Diet, 30 Day 86, 90, 91, 105–81
  beverages 94–5
  breakfast 86, 105–81
  desserts 93–4 see also recipes
  following on 95
  forbidden foods 91
  goals of 92–3
  lunches and dinners see recipes
  portion size 91, 94
  shopping for 89, 91
  snacks 94
  tips 96
  and vegetarians 93
  versatility of 93
  weighing 91–2
dietary goals 92–3
dietary habits 27, 32–34, 63, 85
direction, lack of 78
disaccharides 18–19
drinks
  fizzy 63, 95, 97
  sweet 63–6

eating disorders 82–4
exercise 54, 74, 76, 90
  aerobic 87
  programme 87

fibre 19
food labelling
  COMA report 22
  Food Magazine, The survey 62–3
food lists 96–104
fructose 18, 19, 50
galactose 18
gallstones 24
glucagon 51
glucose 18, 50

hidden sugars 22, 62–4
high blood pressure 24
hyperactivity 26
hypertriglyceridaemia 25
hypoglycaemia (low blood sugar) 49,
  51, 52–3
  reactive 25, 49, 53–4

icecream 13–14
insulin 24, 25, 50, 51, 53, 54, 56
irregular eating 54

'Kreb's Cycle, The' (Hans Kreb) 56

lactose 18–19

magnesium 57, 90

mannitol 70
medical health COMA report 21
molasses 16
monosaccharides 17–18, 50

nutrient-rich foods 91, 98–104
nutritional
  contents
    sugar 44–8
  supplements, 88, 90
    Sugar Factor 58, 88
  training, lack of 22

obesity 21, 24, 54
overweight 73–4, 91

polysaccharides 19
Pre-menstrual syndrome (PMS) 38, 76, 89
  Beat PMT Through Diet 76
  WNAS clinics 76
professional help 36, 37, 73, 77, 88, 89

reactive hypoglycaemia *see* hypoglycaemia
recipes
  chicken
    baked burgers 137
    creamy curry 166
    and parsley rolls 151
    peach and ham-stuffed drumsticks
      160
    with peanut sauce 119
    Spanish 131
    yoghurt roast 156
  desserts
    apricot sherbet 113
    baked apple 120
    baked bananas 109
    blackcurrant sorbet 116
    cinnamon rhubarb 107
    dried fruit compote 157
    frozen peach treat 167
    fruit snow 134
    frumble topping 138
    ginger fruit salad 155
    gooseberry crumble 144
    gooseberry mousse 179
    grape and white wine jelly 174
    jellied grapefruit 178
    nut de la crème 132
    orange jelly 151
    pineapple water-ice 126
    summer pudding 148
  dip
    Greek yoghurt and cucumber
      162
  fish
    baked avocado with tuna 127
    balls 146
    chilli haddock casserole 112
    mustard cod crêpes 172
    prawn-stuffed tomatoes 125
    quick prawn curry 158

salmon steaks with ginger 163
    smoked mackerel pâté 121
    tandoori mackerel 168
    trout with watercress, grape and
      yoghurt dressing 140
  meat
    beef kebabs with mint yoghurt
      153
    beef stir-fry with apricots and
      walnuts 122
    lamb cutlets with nutty apricot and
      mint stuffing 115
    liver and bacon in tomato sauce
      133
    pork with peppers and black
      bean sauce 164
    quick ham risotto 175
  pulses
    red lentil dal 139
  salad dressings
    coconut mayonnaise 118
    Mustard 118
    orange chilli 118
  salads
    apple and nut 124
    beansprout 117
    brown rice and watercress 106
    coleslaw 142
    endive, fruit and nut 128
    Greek 146
    Niçoise 152
    pasta 136
    pepper and lentil 150
    potato 118
    tropical rice 111
  sauces
    chestnut 143
    light cheese 147
    orange 106
    plum and ginger 177
  soups
    avocado and chive 150
    cauliflower 110
    potato and basil 130
    watercress 166
  stock
    home-made chicken 111
  vegetables
    baked avocado with tuna 127
    broccoli with coconut and cashews
      129
    creamy vegetable bake 108
    leek gratin 161
    mixed root 173
    mushroom and spinach yoghurt
      quiche 176
    mushrooms with buckwheat
      stuffing 170
    prawn-stuffed tomatoes 125
    tofu, bean and herb stir-fry 154

saccharin 68–9
smoking 54, 77

snacks
  alternative 94, 123
sorbitol 70
starches 19
stress 51, 77
sucrose
  benefits and hazards 23
  composition 18
  intolerance 25
  loss of nutrients 17
  manufacture 16–17
sugar
  added 22, 24, 25, 28
  advertising *see* advertising
  beet 15, 16
  cane 12, 16
  cereal bars 61–3
  consumption *see* sugar consumption
  craving *see* sugar craving
  definition 17–19
  domestic use of 15
  export 15
  extrinsic 23
  fruit drinks 64–5
  hidden 22, 62–4
  industrial uses 13
  intrinsic 23, 30, 67
  manufacturers 27, 28
  moderate use of 48
  non-milk extrinsic 21, 22
  nutritional value of 44–8
  production 16–17
  purity of 16–17, 45
  sucrose *see* sucrose
  Sugar Craving Syndrome 32–7, 76
  Third World countries 15
  types of 17, 18–19
  uses of 45
sugar consumption, 11–15
  and climate 13–14
  Committee on Medical Aspects of
    Food Policy (COMA) 20–3
  and disease 20–6, 60
  domestic 13
  excessive 45–6
  health
    effects on 20–6, 46, 82–4
  hyperactivity 26
  icecream 13–14
  mood and behaviour
    effects on 26
  National Advisory Committee on
    Nutrition Education (NACNE) 20
  and occupation 40
  statistics 12, 23, 35, 38, 40–3
  survey 40–3
  and tooth decay 20, 21, 26, 31
  United States Food and Drug
    Administration Sugars Task Force
    20, 26, 46
sugar craving
  cost of 42
  men 38–9, 84

  and mental health 82–4
  overcoming 86–9
  and Pre-menstrual Syndrome (PMS)
    38, 76, 89
  professional help 36, 37, 73, 75, 77,
    88, 89
sugar craving *cont.*
  questionnaire 34
  reasons for 32–3, 72–3
  solutions to 33, 73–81
  symptoms of 35–7, 42
  and will-power 92
  withdrawal symptoms 89, 92
  women 38
Sugar Factor 58, 88
sugar-rich foods 46–7
  moderate use of 48
  nutrient content of 46–7
  psychological dependence on 33,
    72–81, 86
  weekly consumption of 41
sweeteners
  artificial 68–71
  disadvantages of 71
  non-nutritive 68–70
  nutritive 70
  side-effects 69, 70

tea drinking 20, 21
tooth decay 20, 21
  causes of 60–6
  cereal bars 61–3
  child dental health survey 59
  children's drinks 64–6
  COMA reports 21, 59
  dental hygiene 61, 66, 86
  and diet 61
  fluoride 60, 61
  hidden sugars 62–4
  intrinsic sugars 67
  plaque 60–1
  prevention 31, 66–7
  protection against 70
  Streptococcus mutans 60
  and sugar consumption 31, 59–67
  Vipeholm study 66
treacle 16

vegetarians 86, 93
Vitality Plan 74, 76

withdrawal symptoms 89, 92
Women's Nutritional Advisory Service
    (WNAS)
  clinics 76, 88
  programme 36, 37, 73
  Sugar Factor 58, 88
  surveys 22, 35, 38, 40

xylitol 70

Yudkin, Professor John
  Pure, White and Deadly 60

If you have enjoyed

Beat Sugar Craving

you may also be interested in the following titles
by the same author:

Beat PMT Through Diet (£6.99)

Beat PMT Cookbook (£6.99)

To obtain your copy, simply telephone Murlyn Services on

0279 427203

Your copy will be dispatched to you without delay,
postage and packing free. You may pay by cheque/postal
order/VISA and should allow 28 days for delivery.